八十二歲白石

Zuo You

The Screens Collected by Zhe-garden

Compiled by The Owner of Zhe-garden

The Forbidden City Publishing House

座右
蔗园藏屏

蔗园主人 编

故宫出版社

图书在版编目（CIP）数据

座右：蔗园藏屏 / 蔗园主人编. -- 北京：故宫出版社，2024.1
ISBN 978-7-5134-1606-1

Ⅰ. ①座… Ⅱ. ①蔗… Ⅲ. ①家具-鉴赏-中国-古代 Ⅳ. ① TS666.202

中国国家版本馆 CIP 数据核字 (2023) 第 234566 号

座 右
蔗园藏屏

蔗园主人 编

顾　　问：	徐　累　张金华
撰　　稿：	张志辉
英文翻译：	付　羽
摄　　影：	山外摄影
责任编辑：	方　妍
装帧设计：	李猛工作室
责任印制：	常晓辉　顾从辉
出版发行：	故宫出版社
	地址：北京市东城区景山前街 4 号　邮编：100009
	电话：010-85007800　010-85007817
	邮箱：ggcb@culturefc.cn
印　　刷：	北京雅昌艺术印刷有限公司
开　　本：	787 毫米 ×1092 毫米　1/8
印　　张：	43
印　　数：	1～1600 册
版　　次：	2024 年 1 月第 1 版
	2024 年 1 月第 1 次印刷
书　　号：	ISBN 978-7-5134-1606-1
定　　价：	890.00 元

目录
CONTENTS

吟诗对画屏 / 邹静之	17
座右——中国古代座屏概谈 / 张志辉	23
图版	59
长河遗珠	61
山水寄情	141
繁华竞逐	273
跋 / 蔗园主人	317
图版索引	324
附录：检测报告	332

RECITING POETRY BEHIND THE SCREEN / Zou Jingzhi	20
SEAT RIGHT: A Brief Talk on Ancient Chinese Seating Screen / Zhang Zhihui	56
PLATES	59
TREASURES FROM THE LONG RIVER OF HISTORY	61
LANDSCAPE SUSTAINS ASPIRATION	141
TIMES OF CHASING PROSPERITY	273
POSTSCRIPT / The Owner of Zhe-garden	321
INDEX	324
APPENDIX: TEST REPORT	332

吟诗对画屏

邹静之

今年9月,我已年过古稀。

七十岁生日过后,人无端地松弛下来,能干什么,不能干什么,心里盘算过了。

出生到现在,所作所为无非是向着一个念头去的——与什么样的人、在什么样的环境下,了此一生。说得具体点:生而为人,所求无非是想与温良恭俭让的人,在山清水秀的环境中过一辈子。

所谓环境,有天造地设,也有人为。

家居的营造就是人为,比如房舍、院落、家具、摆设等。

时代变革,人为的环境也跟着变了,有的物件留下来,有的消失了。古时多见,现已不多见的屏风系列就是个例子。

"屏"这个字,有遮挡、隐蔽、分隔等实用意义,之后才有了礼仪、风水等形而上的说法。

屏在建筑中有照壁、影壁、萧墙等;家具中有画屏、案屏、镜屏、砚屏、灯屏等。

少时读苏东坡,有两句有关屏风的诗,因其画面生动而特别能记住:"倦仆触屏风,饥鼯嗅空案。"(苏轼《除夜病中赠段屯田》)病中的苏学士深夜倦卧,童仆不意间触到了遮护自己睡姿的屏风,惊醒了苏学士,睁眼时,见到了一只饥饿的小鼠,孤灯下正用鼻子嗅着空无一物的书案,一时人鼠两望,道出无尽的凄凉。

这里的屏风,说的是枕屏。

当年古人为了让不雅的睡姿,不被人看了去,倦卧打盹时,要用一架枕屏把自己的身体遮掩起来,这举动说小了是讲究,说大了是文明。再想那些为佳人更衣而设的折屏;《韩熙载夜宴图》中,一对青年男女窃窃私语时相隔的立屏;或为表现尊严在座椅后设的围屏;观书时的灯屏等,一时让人悟出屏风这器物就是人类文明中的一个符号。

当屏风过渡到只为了陈设用的座屏时,实用性似乎就没有装饰性大了。或只为造境,为了审美,为借物移情,在小小的文房中设了一架亲切的座屏,说白了就是为了看着高兴。

衣食无忧后,精神愉悦是人生之大事。

收藏界对陈设品有句行话——越是没用的东西越值钱,盛鱼的盘子卖不过画鱼的盘子。

座屏或插屏，作为登堂入室的观赏陈设器，其出现，还是要有长时间的社会安定繁荣为条件的。古董一类大致如此，比如同为一朝的瓷器，也是盛世时的作品更珍贵些。

再有座屏的出现，与文人有关。

北宋诗人张耒的《慈湖中流遇大风舟危甚食时风止游灵岩寺》中有"从今要见庐山面，画作屏风静处看"的句子。说的是人走了景带不走，只好画于屏上，置于案头静静观看。再读李白的《清溪行》"人行明镜中，鸟度屏风里"，诗仙在清溪岸边走，倒影在溪水中并行，犹如幻境，倏忽间又见鸟儿从画屏中飞过。这种现实与幻象的折叠出现；实与虚，家居与风景的对切，借物咏物，一镜一屏，道出多少意趣啊！

蔗园主人，经数十年潜心收藏，集合了明清以来各式各样的珍贵座屏，实属罕见。其中以天然石材为屏，黄花梨、紫檀作架的座屏、插屏，式样繁多又各有姿态。

这类以石纹笔触、石晕渲染而造出的远山近水，浩瀚苍穹，倘若久看，会有出神之感。

有案头工作经验的同道，如写作，或画画，再或者就是解题作业，都会有神思不至、下不了笔的时候。此时若看着眼前的山河座屏，就会有移情通灵的作用，所谓："相看两不厌，唯有敬亭山。"（李白《独坐敬亭山》）看着看着，心绪一动，文墨落下，也就写出来了。

案上座屏，说它是文人之需、灵感之窗也不为过。

20世纪90年代，我写作之余，常去北京高碑店、吕家营一带旧家具市场游荡，买过两架座屏式的帽镜，一架黄花梨贴皮的，一架广作红木的。那时偶然会看到一些明清大漆螺钿或黄花梨云石的座屏出现，也听来了一点皮毛知识，比如：从立牙子来看，抱鼓墩的比透雕花的年份早；若是大漆的看是不是披麻灰的，看断纹的程度。

为什么没有收？一是因其卖价高；二是自认知度不够；还有非常自我的说辞是：这座屏一请回家，随着它，我那些案子字画都得换了才行，能力有限。以至一些高古的珍品都错过了。

蔗园主人几十年来收藏了这么多将要消失的珍贵座屏，能力是其一，认知高蹈才是最重要的。古董收藏，有心没力不行，有力没心也不成，心力都到了才能有高度。

藏品中有一架明代黑漆薄螺钿孔子观欹器图座屏，尺寸大、人物多、制作精美外，那件欹器的具体呈现，为我解开了多年不知其物为何物的疑惑。两千多年前"谦受益，满招损"的道理就是通过这件器物讲出来了。

这样的一架座屏，当置于文庙的殿堂之上才合适吧！

螺钿镶嵌工艺，考古发现西周已经有了，而在北宋时又出现了薄螺钿的工艺。

古人在诗歌上怎样表现螺钿，查到最早的诗句是元代诗人尹廷高的："蟠螭金函五色毯，钿螺椅子象牙床。"（螺钿在这写成钿螺或因平仄的关系）这两行诗读着平白，但可贵处是非常具体地描绘出了元代家居设置的华丽景象。

到了清代，诗人刘应宾就写出了"螺钿妆成翡翠光"的诗句了。可以看出元明清螺钿镶嵌的家具已是时尚了。

追溯到唐代，镶嵌类的家具，在诗中用得更多的是"云母"而不是"螺钿"。尤其用在屏风上更加如此。刘禹锡写过："云母屏风即施设，可怜荣耀冠当时。"卢纶写过："水精如意刁金色，云母屏风透掩光。"更有李商隐的《嫦娥》写出："云母屏风烛影深，长河渐落晓星沉。"三位唐朝诗人都是用"云母屏风"这个固定的词来描述了当时的镶嵌工艺。

"云母"是一种可以剥的很薄的矿石，剥过后有半透明感，并会发出五色豪光。直到现代的百宝嵌工艺中也会有云母的出现。据此或可推断，在薄螺钿工艺之前的唐代，更多以云母石为主材，加上错金银等工艺来装饰屏风吧。同是唐代诗人温庭筠的名句"小山重叠金明灭"说的就是这类所谓镶嵌工艺的屏风。

老话说：人过百岁既为"人瑞"，物过百年便是"祥瑞"。

蔗园主人几十年来，集结了如此丰富的祥瑞珍宝，能力除外，实实在在地有一颗尊重文明的心。

以心聚物，现今又出版了《座右》这部大书，将所藏宝物昭告天下，广布人间，功德无量！

2022 年秋

Reciting Poetry behind the Screen

Zou Jingzhi

The owner of the Zhe-garden, after decades of collecting and studying, has gathered a rare collection of various precious screens from the Ming and Qing dynasties. The screens and inserts made of natural stones and made of *huanghuali* and red sandalwood frames are diverse in style.

These screens, created with stone texture rendering, depict distant mountains and rivers, vast sky, and if gazed at for a long time, can evoke a feeling of transcendence. For those who have experience in such tasks as writing, drawing, or solving problems, there are times when they may experience a mental block and cannot put their thoughts down on paper. As they gaze upon the screen, their minds may be inspired, and their writing or painting may flow more freely.

The screens on the desk are not only necessary for literary scholars but also serve as a source for inspiration.

In the 1990s, while writing, I often wandered around the old furniture markets of Gaobeidian and Lujiaying in Beijing. I once bought two hat mirrors in the form of screens, one made of *huanghuali* and the other made of *hongmu*.

Occasionally, I would come across some Ming and Qing lacquered screens with mother-of-pearl or *huanghuali* and marble, and I learned such facts as the screens with drum shaped footings are older and whether the lacquer has signs of hemp and grey ashes underneath the lacquer layer and the degree of surface crazing.

Why didn't I buy them? One reason was that they were too expensive, another was that I didn't feel confident in my knowledge, and there was also a very personal reason: if I brought the screen home, all my desk items and paintings would have to be changed, and my resources were limited. As a result, some of the more valuable antiques were missed.

For decades, the owner of the Zhe-garden had collected so many precious screens which are becoming rarer. Finance is one thing, but knowledge and discernment are the most important. Collecting antiques requires both passion and resources. Only when both are present can one achieve a higher level.

Among the collection is a large Ming lacquered screen inlaid with mother-of-pearl called 'Confucius Observing the Tilting Vessel', which not only has a large size and many figures but also has exquisite craftsmanship. This screen depicts the specific presentation of the tilted vessel, which finally helped me understand the meaning of 'Humility will bring benefit; arrogance will bring harm' which was spoken about more than two thousand years ago. Such a screen would be best placed in the hall of a Confucian temple.

In archaeological findings, inlaying with mother-of-pearl technology was already present in the Western Zhou Dynasty. During the Northern Song Dynasty, the technology of thin inlay with mother-of-pearl emerged.

In the Tang Dynasty, lacquer inlay with mother-of-pearl method was used as a decorative technique on furniture. In poetry, this technique was described as 'Yunmu' but not 'Luodian' especially when this method appeared on screen. The term 'Yunmu' was used to described mother-of-pearl inlay techniques at that time appearing in the poetry of Liu Yuxi, Lu Lun and Li Shangyin.

'Yunmu' is a kind of mineral that can be peeled very thinly, with a semi-transparent quality and emits a magnificent five-colour light. Even in modern jewellery inlaying crafts, mica can still be used. Based on this, it can be inferred that in the Tang Dynasty before the thin mother-of-pearl craft emerged, mica was the main material used for decorating screens, along with techniques such as gold and silver inlaying. The famous line by Tang poet Wen Tingyun, 'The small mountain overlaps and the gold shines and dims,' refers to this type of inlaying craft used in screens.

There's an old saying that when a person lives to be over 100 years old, he is considered a 'centenarian', and when an object reaches over 100 years old, it is considered a 'treasured omen'. For decades, the owner of the Zhe-garden has gathered such a wealth of precious objects, and apart from his talent, he has a great passion for respecting civilization. By gathering these objects with his passion, he has now published a comprehensive book called *Zuoyou,* which announces his collection to the world and distributes it widely, extending our knowledge and advancing the culture of all mankind.

座右
中国古代座屏概谈

张志辉

屏具是我国家具门类中甚为特殊的一类，历史悠久、脉络清楚、独成体系。屏具原因礼法而设，是文明发展到一定阶段的产物，并非日常生活所必需，故在家具中它的出现比席、床、案等略晚，但至迟周代已有。屏具的装饰、隔断、遮蔽等功能，是随着历史发展逐渐附加的。

"屏"字原意并非指屏具。郑玄注《礼记》曰："屏谓之树，树所以蔽行道，管氏树塞门，塞犹蔽也，礼，天子外屏，诸侯内屏……"[1]"树塞门"，是建筑前以树木起屏蔽、遮挡作用，更接近后世院落中的照壁。"屏"字含义包括屏具，即"屏风"的出现，当在战国时期，但均见于西汉典籍，"孟尝君待客坐语，而屏风后常有侍史"[2]，即是。

根据造型，屏可分座屏和围屏两大类。顾名思义，座屏即屏下附有底座者，这是屏具中最先出现的类型，周时已有。座屏起初为大型或中型屏，陈设于地，多呈"一"字形，如长沙马王堆一号墓出土彩绘座屏模型[3]〔图1〕，就是非常有代表性的早期座屏。所见五代及以前座屏形象，是在屏扇下附有两个独立的屏墩固定。至迟北宋时，就出现了两个屏墩以横枨相连的屏座，也出现了站牙（或连弧形斜枨）、抱鼓墩、披水牙板、绦环板等构件，座屏造型已经成熟。此时的座屏屏扇大多与屏座一体，也有分体者，则是屏墩上设立柱，立柱有槽口，可以卡住屏扇，这属于座屏中的插屏，俗称"插牌子"。

围屏是一种多扇式屏，以铰链等相连，可以不同角度作围合状，又名曲屏。在某些历史时段或某些样式中，座屏与围屏互有交融，本文亦会顺便述及围屏的相关问题，但并不特意梳理围屏的脉络。

本文关心和讨论的对象为屏具中的座屏，以朝代为序，介绍其发展的大概历程。需要阐明的是朝代只是作为特定时间区间来看待，政权更替或对座屏在内的器具发展有所影响，但不是其变化的必然节点。另，座屏的核心为屏心的内容，其他龙骨、屏框和屏座是作为附属而存在的，但从家具史研究的角度，两者都关心，甚至后者更多一些，本文亦是。

1 西汉 彩绘座屏模型（线图）
马王堆一号汉墓出土

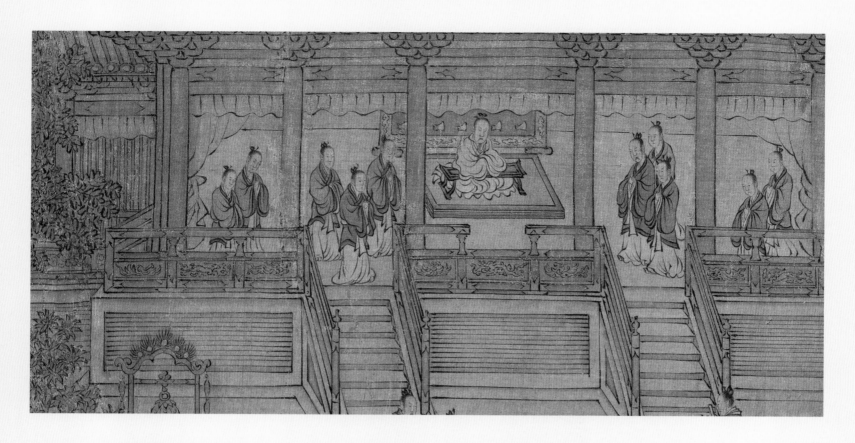

2 | 宋 马和之《小雅鹿鸣之什》卷局部
故宫博物院藏

一 战国及以前的屏具

周时的屏具，有黼依和皇邸。又有依、扆、斧依、黼扆、斧扆等，应与黼依同指一物。《礼记》"曲礼"载周天子"当依而立"〔4〕。郑玄注曰："依，状如屏风，以绛为质，高八尺，东西当户牖之间，绣为斧纹也，亦曰'斧依'。"〔5〕《周礼》"春官宗伯"亦记："凡大朝觐、大飨射，凡封国、命诸侯，王位设黼依，依前南向。"〔6〕

汉时八尺合今184厘米左右，超过一般人高度。黼纹是王者的象征，高诱等认为"白与黑谓之黼"〔7〕，颜师古等则认为"白与黑画为斧形谓之黼"〔8〕。唐友波先生通过战国青铜春成侯盉和长子盉上纹饰与金文的关联性研究，并对各家之前关于黼纹的观点进行总结，提出黼纹应为"勾连雷纹"〔9〕。依通"扆"，"户牖之间谓之扆"〔10〕。天子南面，黼依陈设位置就应在明堂之北，是彰显天子权力、地位的器具。这一方式一直为后世所继承，明清时期的殿堂，在宝座后陈设大座屏，都是黼依余绪。

宋人复原黼依图像，可见马和之《小雅鹿鸣之什》〔图2〕和聂崇义《新定三礼图》〔图3〕。前者"彤弓"部分绘周天子赏赐诸侯并设飨之事，明堂上天子面南席地而坐，诸侯列于两旁，在两柱间北侧天子后方，设座屏。座屏绛色地上饰以斧纹，有宽边框，下设屏墩。除了将黼纹认为是"斧纹"值得商榷外，就造型来看，片状屏扇下附以屏座，是屏具最基本、最原始的样式，当与黼依相差不远。马王堆西汉墓出土的两件彩绘座屏模型，亦是类似样式（见图1）。

皇邸是另一种早期屏具。《周礼》"天官冢宰"："掌次，王大旅上帝，则张毡案，设皇邸。"郑玄注为："皇，羽覆上，邸，后版也。"[11] 皇邸是饰以凤凰羽毛（即美丽的鸟羽）的板状屏具。皇邸是天子祭祀上帝专设，是上帝位置物化的一部分，方位上是天子所面对者，与其面南时所背的黼依完全不同。皇邸参与上帝之位的确立，黼依参与皇帝之位的确立。皇邸后世不传，今已不知其形象，但设祭时案后设屏的组合方式，一直流传，明清时亦然[12]。

战国及以前的屏具图像和实例极少[13]。战国中山王䂮墓出土有L形曲屏底座及残片。该墓出土铜错金银虎噬鹿屏座、铜错金银犀屏座、铜错金银牛屏座一套[14]，铜错金银虎噬鹿屏座施于屏的转角处，上有銎口，可使屏成84度交角状。铜错金银犀屏座、铜错金银牛屏座分别施于两扇屏的外下端。漆屏心夹纻胎，屏框黑漆地上饰朱、橙色花纹，可辨者有鸟纹、涡纹等。横框长约2米，竖框长约1.1米，复原后为一高约1.3米，两向长约2米的L形曲屏。此外还出土有铜合页、辅首等附件，合页可折叠，屏不设时可拆下折叠搁置，出土时的情况亦是如此[15]。从造型来看，战国中山王䂮墓出土的曲屏介于座屏和围屏之间。

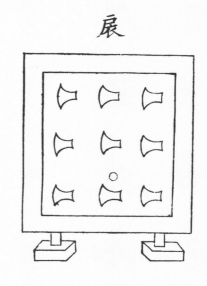

3 | 宋 聂崇义《新定三礼图》插图
清康熙十二年通志堂影刻本

二　秦、汉时期的屏具

秦时屏具相关资料虽少，历史事件荆轲刺秦中，姬人发琴音："罗縠单衣，可掣而绝。八尺屏风，可超而越。鹿卢之剑，可负而拔。"[16] 这是有关秦人使用屏风的记载。从战国时已屡见记载和汉时屏具的广泛应用来看，秦时的屏具应该有更进一步发展。

汉代是自春秋战国以来大一统时间甚久的朝代，政治、经济、文化等方面都有长足的发展，工艺进步，中国家具史上低型家具体系在此时达到高峰。汉代家具门类丰富，功能齐全，装饰繁华瑰丽，造型多变，屏具极大丰富。《西京杂记》载："赵飞燕女弟居昭阳殿，中庭彤朱，而殿上丹漆……中设木画屏风，文如蜘蛛丝缕。"[17] 又载赵飞燕女弟在昭阳殿遗书，有云母屏风、琉璃屏风[18]。汉武帝"为七宝床、杂宝案、厕宝屏风、列宝帐，设于桂宫，时人谓之四宝宫"[19]。中山王为鲁恭王文木作赋，文木可"制为屏风，郁岪穹隆"[20]。"董偃常卧延清之室，以画石为床……上设紫琉璃帐、火齐屏风。"[21] 都是以珍奇材料制作屏具的记载。"一杯棬用百人之力，一屏风就万人之功。"[22] 屏具已不仅是礼仪、屏蔽之器，也成了宣示权力、地位的奢侈之具。此时已有施于床的屏，西汉时陈万年教诫儿子陈咸于床下，"语至夜半，咸睡，头触屏风"[23]。

汉时的屏具还有隔坐功能。东汉时，郑弘"代邓彪为太尉时，举将第五伦为司空，班次在下。每正朔朝见，弘曲躬而自卑，帝问知其故，遂听置云母屏风分隔其间"[24]。这是以屏风隔坐，解除朝臣间的尴尬，类似故事还见于三国吴时纪陟[25]。这类屏具或称为"隔坐屏风"，或直接称为"隔坐"。

汉时屏具主要有"一"字形座屏和L形曲屏两种。"一"字形座屏形象，即矩形屏板一块，附屏墩以固定，这是屏具的最基本形态。汉代座屏实例以西汉马王堆一号墓出土的彩绘座屏模型最为典型[26]〔图4〕。座屏通高62厘米，屏板长72厘米、宽58厘米、厚2.5厘米，制作较粗糙，应为陪葬明器。正面饰穿璧纹[27]，周匝有变形勾连纹饰，背面饰云龙纹，正背两面皆饰有锦纹边框。屏墩近鱼尾式造型，为汉时家具常见腿足或枕足造型。该墓同出土的遣册上有"木五菜（彩）画并（屏）风一，长五尺，高三尺"字样，即指此类屏。此屏所在的北边箱出土时器物次序井然，摆放有度：座屏位于最西首，周匝有帷幔、木杖、漆奁、香囊、枕巾、漆凭几等，前有食案，上列漆盘、漆耳杯、漆卮、竹箸、竹串等。食案前有各式乐舞俑和侍俑，显然这是一个宴乐场景的真实反映[28]。屏处于东西向中轴线的最后端，设在主人位置之后，与黼依之制相合。湖南长沙马王堆三号汉墓、甘肃武威东汉墓、河南洛阳涧西七里河东汉墓都曾出土类似造型的座屏或座屏模型[29]。东汉彩绘孝子故事漆奁[30]、内蒙古和林格尔县小板申村东汉壁画墓《拜谒、百戏图》[31]、辽阳三道壕窑业第二现场东汉令张君壁画墓《家居图》[32]、沂南东汉晚期墓画像石《卫姬谏齐桓公故事图》[33]等，都可见使用这类座屏的场景。这种座屏样式为中国座屏之经典，后世除了屏墩更加复杂，增设站牙、牙板（或披水牙板）等辅助性构件外，无大的变化。

L形曲屏仍然流行，分两种形式，一种延续前述中山王䓨墓带座形式，介于座屏和围屏之间。西汉南越王墓出土一组大座屏，复原后高1.8米、正面宽3米、两

4 | 西汉 彩绘座屏模型
马王堆一号墓出土

侧各宽1米，正面分三部分，中间为向后开启的两扇门，两侧屏扇与背面屏扇间有可折叠的铜构件，制作极精巧[34]。此座屏外形和魏晋时流行的"冂"形三面围屏（以下以"三面围屏"专指）相似，然其造型未尝不可看作两个L形曲屏和门扇的组合式屏，其门扇底框位置较低，明显区别于两侧曲屏，应也是有意为之。另一种曲屏直接施于席或床榻，更近围屏，基本造型即两扇等高的屏连以合页，用时打开近直角，两扇相抵而立，多有装饰纹饰的边框〔图5〕。颜师古所谓黼依"形如屏风而曲之"[35]。羊胜《屏风赋》中"屏风鞈匝，蔽我君王"句[36]，都是对这种曲屏的描写[37]。

L形曲屏是汉或以前独特的一类屏具，并非中轴线对称的方式值得注意，这与包括屏具在内的大部分传统器具造型布局方式迥异。L形曲屏在魏晋时渐被三面围屏取代而式微，后再无所见，亦未为任何家具造型继承或借鉴[38]。

汉代屏具于室内陈设时，已因尊卑有造型、尺度的变化，山东安丘市王封村东汉画像石《拜谒图》中夫妻所用之屏大小迥异，三道壕窑业第四现场《家居饮食图》中夫妻所用之屏高低不同，三道壕窑业第二现场《家居图》中张君背设直屏而妇人不设屏，都是这种体现。

汉时屏具的装饰，除了常见的穿璧纹、勾连纹、锦纹、云龙纹、忍冬纹、云气纹、鸟兽纹等装饰图案外，先贤、列女等人物图画也开始成为屏具上的主题，屏具除礼仪、装饰、隔断等功能外，尚有劝诫、教化功能，刘向"与黄门侍郎歆以《列女传》种类相从为七篇，以著祸福荣辱之效，是非得失之分，画之于屏风四堵"[39]。羊胜《屏风赋》有"重葩累绣，沓璧连璋。饰以文锦，映以流黄。画以古列，颙颙昂昂"句[40]。也是以明君圣贤等为主题的屏具。此类屏具的兴盛，也与画学中人物画之发展有关。东汉时，"宋宏尝燕见，御座新施屏风，图画列女，世祖数顾视之，宏曰：'未见好德如好色者。'帝撤之"[41]。亦侧面反映屏风画的生动传神。

5 | 东汉 画像石《拜谒图》
山东省安丘市王封村出土

三 三国、魏晋、南北朝时期的屏具

此时北方和西北民族内迁，政权更迭频繁，外来文化与中原文化相互碰撞、交融。汉时传入的佛教，在这一时期已经甚为普及。这些都对这时期的家具发展产生了深远影响。交杌、绳床、筌蹄等适合垂足而坐的坐具进入中原，中国家具中适应席地而坐的低型家具体系渐衰，垂足而坐的高型家具体系渐得以发展。

三国时吴少帝"孙亮作流离屏风，镂作《瑞应图》，凡一百二十种"[42]。后赵武帝石虎有"金银钮屈戌屏风，衣以白缣，画义士、仙人、禽兽之像，赞者皆三十二言，高施则八尺，下施四尺，或施六尺，随意所欲也"[43]。屏具开始成为图画的重要载体。

魏晋时期座屏并未有更新的发展，围屏是屏具中的主流，既有三扇式的三面围屏，也有多扇式围屏。魏晋时期开始流行三面围屏，以传顾恺之《列女仁智图》中所见和司马金龙墓漆屏图画最为典型。

传顾恺之《列女仁智图》（宋摹本，故宫博物院藏）中的三面围屏是反映施于地的用法〔图6〕，其上的山水画主题已被临摹者改为宋时山水[44]，但围屏的样式和使用方式还延续魏晋样式。图中还出现了一个小三面围屏遮蔽座灯，是灯屏的早期形式。济南东魏崔令婆墓出土滑石柱础四件，长11厘米、高5.5厘米，其中两件杀掉四分之一，应是一种施于这种三面围屏的两外转角处的屏墩[45]。这种用石或其他比较沉重材质做成屏墩、固定围屏的做法，一直到明清尚有沿用，多呈鼓形。

司马金龙墓出土的木板彩漆屏[46]〔图7〕，因遭盗掘被破坏，部分留存，为长约80厘米、宽约20厘米、厚2.5厘米的彩画木板组成，各屏板件有暗销相连，外缘有榫舌。屏红漆为地，墨线勾勒人物故事图案，人物面、手部涂铅白，其余有黄、青绿（深浅不同）、橙红、灰蓝等色。人物皆有标记，每组故事旁有题记，皆是黄地墨字。屏上下四栏，横向各屏相连为通栏式，一面为列女故事，一面为帝王先贤故事，劝诫意图明显[47]。图画与传顾恺之《女史箴图》《列女仁智图》等风格、构图相近，是当时流行之图式。同墓出土的还有漆画边框，宽7厘米、厚5厘米、

6 　(传) 东晋 顾恺之《列女仁智图》卷局部
　　故宫博物院藏

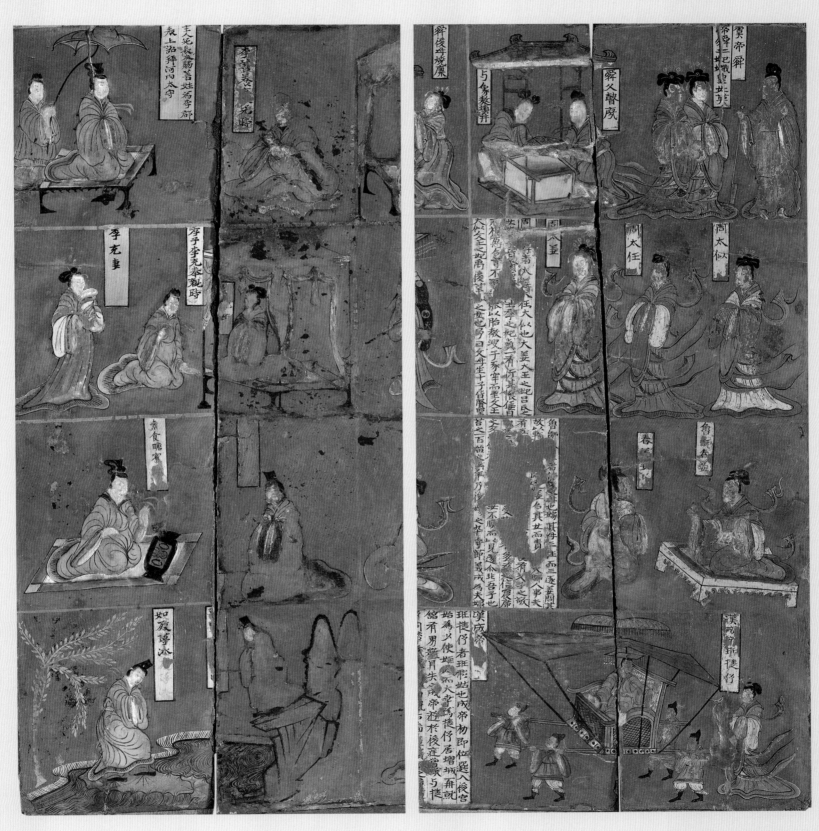

7 | 北魏 木板彩漆屏
司马金龙墓出土

8 | 北魏 木板彩漆屏上的三面围屏小榻
司马金龙墓出土

长 15～85.2 厘米不等，上饰青龙、白虎、朱雀、鹿形兽、小鸟、行云、童子等，盘绕忍冬花纹。边框一侧有槽口，可容纳屏心上的榫舌。此屏因损坏不知原貌，难得的是屏上绘有两件设围屏独坐小榻，一则可知当时这种三面围屏在上层社会的流行；二则未尝不是复原漆屏的参考。围屏小榻出现在《卫灵公与卫灵公夫人》及《和帝口后》（图8）两处，后者画面保存较完整，形象表达肯定。图绘女眷（当为口后）坐于方榻，上设内为网格纹屏心、外有双边框的三面围屏。以此为参考，假定漆屏原有 12 片之多[48]，可将漆屏复原为高约 94 厘米的三面围屏，背面 6 片屏板宽约 130 厘米，侧面各三宽屏板宽约 70 厘米，使用时或施于榻上[49]。

魏晋时的 L 形曲屏和三面围屏较多见一些网格纹屏心，如司马金龙墓漆屏所见，此外还出现了一种鱼鳞状屏心，见于北魏洛阳石棺床线刻画像[50]和北魏大同智家堡石椁墓壁画等[51]。

魏晋南北朝时期新出现一种多扇围屏，在当时甚为流行。这类围屏的图画已经从之前诸如司马金龙墓的横向展开式转化为竖向条屏式，每扇屏都是一个独立的图案单元，折叠更为灵便。传顾恺之《女史箴图》（宋摹本，大英博物馆藏）之《夫妻并坐图》中，绘一壸门大床，上施帷帐，前设栅足长矮几，床之上围以多扇屏，除了三面围合外，正面还有两扇合围，两扇半开者[52]（图9）。屏扇宽边框，屏心为网格纹，屏扇间以外红内黑线条表示织物裱糊成的合页。从女眷可架腋屏上看，屏高当在 60 厘米左右。此图所表现的床、围屏、帷帐的组合方式，是后世架子床之雏形[53]。山西大同御东建设区北魏太安四年（458 年）解兴墓壁画《墓主图》中夫妻并坐，后方屏心网格状的多扇屏，高及坐者耳部[54]。其他如山西太原北齐徐显秀墓[55]、济南北朝东八里洼墓壁画[56]所见的多扇围屏榻都甚为典型。

石围屏榻是北魏时期新出现的一种葬具，其常见形象为六足或四足壸门石榻，三面石围屏，上浅浮雕数组图案，以边框界开，示意为多扇屏组合而成，多浮雕人物故事图，整体图案显然已见佛教和胡人文化的影响印记。如分藏于日本奈良天理大学附属天理参考馆和美国旧金山亚洲艺术博物馆的一套石围屏，由 12 幅人物图案组成[57]。此套石围屏图案还出现围屏形象，其中《墓主夫妇》主题表现夫妇二人坐于壸门大榻之上，三面围以多扇围屏，屏高恰足以遮蔽坐于其上的人，侧面两扇屏折为 Z 形，折合方式与《女史箴图》相类。在西安北周康业墓[58]、西安北郊北周安伽墓[59]等则出现一种装饰有粟特胡人祆教图案特征的石围屏榻，这种葬习一直延续到隋唐时期。

石围屏榻是一种极为特殊的葬具，应该在当时存在过相似造型和装饰风格的实用围屏榻。虽不能将这种依附于榻的石围屏完全看作是屏具的一种，但可将之视作围屏源流之一，并借此来考察围屏的形制。屏之间的连接方式，概与司马金龙漆屏类似，同一侧屏之间以暗销连为一体，两侧之间有屈戌或合页等构件，可以折叠。石围屏榻上多扇围屏的样式，开唐代六曲屏风之先，亦是明清围屏的前身[60]。

9 | （传）东晋 顾恺之《女史箴图》卷局部
大英博物馆藏

四　隋、唐、五代时期的屏具

隋唐时期是高型家具大发展的时代，垂足而坐的方式开始流行，高型家具体系大发展，椅子（时作"倚子"）、桌（时作"卓"）出现，低型坐具如席已经渐渐退出历史舞台。石围屏榻依然见于隋唐时期的墓葬，诸如甘肃天水市石马坪文山顶隋唐墓就出土一件[61]。日常生活中，这种多扇围屏发展成为六曲屏风（兼有少量八曲、十曲屏风），见于唐人诗词，证诸佛窟壁画、墓室壁画和少量的实物留存。

五代上继唐，下启北宋，是家具史上甚为重要的时期，壸门结构衍化而成的插肩榫式家具开始出现（以鹤膝桌和云纹腿榻为典型），日用坐具从坐榻开始往椅子转换，屏具中"一"字形大座屏（即高度在一人高以上者，后同）和"⌒"形三扇式大座屏开始广泛应用。

唐李贺《屏风曲》："蝶栖石竹银交关，水凝绿鸭瑠璃钱。团回六曲抱膏兰，将鬟镜上掷金蝉。沉香火暖茱萸烟，酒觥绾带新承怀。月风吹露屏外寒，城上乌啼楚女眠。"[62]描述了一套饰有蝴蝶石竹，装有银交关（即合页）的六曲围屏。李商隐亦有"六曲连环接翠帷"句[63]。六曲屏风，即一种六扇屏以合页或铰链、丝绳等连成，可以弯曲折叠，互成夹角而立，是唐人生活中甚为常见的围屏样式。敦煌唐代佛窟壁画、唐墓壁画，都可见到丰富的六曲屏风形象。

敦煌唐代壁画中，六曲屏风应用最典型的当属《维摩诘经变》主题壁画[64]。如盛唐第103窟所见，维摩诘坐于榻上，前有栅足凭几；旁有高柱，支撑帷帐；周匝围有高过头顶的六曲屏风，后方四扇，两侧各折一扇。每扇屏心竖六横三分为数格，底色以赭石色和白色两种交叉，上均为两行草书题字（多为六字），是一件精彩的书法屏风〔图10〕。《历代名画记》记定水寺"殿内东壁孙尚子画维摩诘，其后屏风临古迹帖亦妙"，即是此类。六曲屏风的身影在各式表现室内场景的敦煌壁画中，亦是司空见惯，图像多是设壶门大榻，榻后围以六曲屏风，形式大同小异，装饰图案既有团窠纹等，亦有花鸟、山水图案等[65]。

唐代墓室壁画所见的六曲屏风较多[66]，如陕西长安县王村墓仕女图六扇屏[67]、陕西富平吕村乡朱家道李道坚墓山水画六扇屏[68]、新疆阿斯塔那217号唐墓的花鸟六扇屏[69]、216号墓的《鉴戒图》六扇屏[70]等，其中《鉴戒图》中出现了敧器、石人、金人、土人、扑满、丝束、青草等有鉴戒寓意的图像，而本书所录明黑漆螺钿孔子观敧器图座屏正是这种鉴戒图的遗意。还有一种树下老人题材的六扇屏亦较为常见，如山西太原南郊金胜村四号墓、五号墓[71]、六号墓[72]、焦化厂墓[73]皆有所见。墓葬所见的唐代六曲屏风，除了两侧各一扇前折的形式外，还有各三扇成L形者、六扇平列者[74]。

唐代六曲屏风实物，传世者见于日本奈良东大寺正仓院，虽有些不排除为仿唐之作，但亦可借此了解唐代屏风状况。正仓院保存有日本光明皇后献给东大寺的圣武天皇遗物（相当于唐天宝晚期）。据当时献物账记载，有屏风106扇，今所存者有六曲屏风2架和27扇残缺者。完整者一为"鸟毛立女屏风"，图案为树下唐装仕女，姿态各不相同，面貌趋近，据传仕女衣服和树叶上原粘鸟毛装饰（现只有一扇上仕女袖间余少许）。另一套完整者为"鸟毛帖成文书屏风"。此外还有残缺的鸟毛篆书屏风、夹缬屏风等，题材有人物、山水、鸟、兽等，这些屏高度多为148~163厘米、宽度多为54~56厘米，装饰题材和尺寸与前述敦煌壁画和墓室壁画中所见相合。正仓院所藏这批六曲屏风，从"鸟毛帖成文书屏风"和"鸟毛篆书屏风"边框上的痕迹看，是在近两端处有丝绳或其他软物穿过为铰链，可以前后折叠。我国新疆阿斯塔那地区唐墓曾出土的六曲屏风残件[75]，皆是木框曲屏、绢上作画。其中187号墓屏风绘盛装妇女对弈，并有侍婢、儿童等，230号墓屏风上裱绛紫绫边，绘略具舞姿舞蹈或持乐的仕女，188号墓屏风图案以牧马为题。2019年，甘肃武威武周吐谷浑喜王慕容智墓出土有围屏，长一米余，木质框架尚在，呈"目"字形，上裱织物，有精细的花卉图案，甚为罕见[76]。

相比于丰富的曲屏资料，唐代大座屏的应用情况所知甚少。五代时图像中见有一种大型绕床（或榻）座屏，或许唐时已出现。北京市海淀区八里庄晚唐墓北侧有

10 右页图
唐《维摩诘经变》壁画
敦煌莫高窟103窟东壁

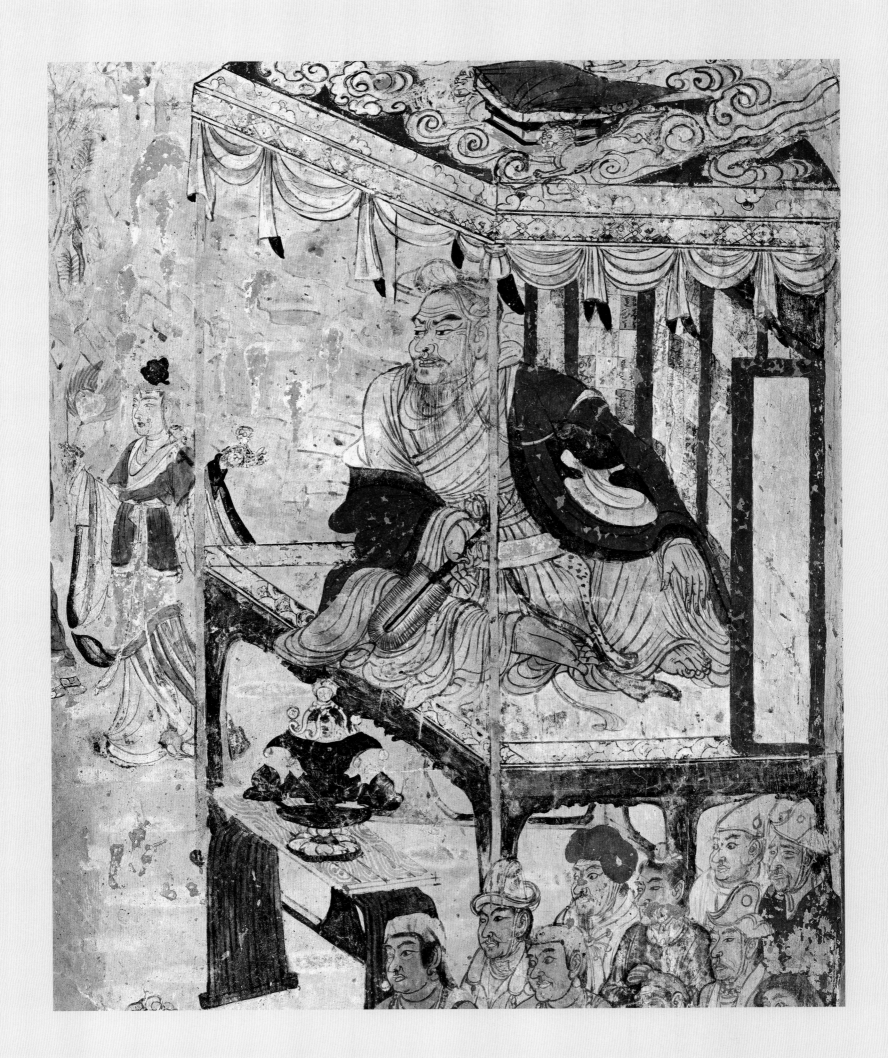

棺床，其旁弧形壁上绘一幅长290厘米、宽156厘米通景画，周匝为宽约11厘米的朱红色边框，中绘一株硕大的牡丹，周匝点缀芦雁、彩蝶、蜀葵、百合等[77]，此幅壁画无论从布局形式还是内容，都可见床旁花鸟大座屏的端倪。

唐代六曲屏风的盛行，与屏风书画的兴盛不无关系，盖因曲屏方便摆放、挪动，收展自如，糊以纸、绢等，便于书画作品的展示，或为教戒目的，或为装点室内，或为抒发情感，无所不可。一时屏风书画名手迭出，白居易《素屏谣》提到"李阳冰之篆字，张旭之笔迹，边鸾之花鸟，张藻之松石"[78]。杜牧诗《屏风绝句》："屏风周昉画纤腰，岁久丹青色半销。斜倚玉窗鸾发女，拂尘犹自妒娇娆。"[79]《历代名画记》记薛稷"尤善花鸟、人物、杂画，画鹤知名，屏风六扇鹤样，自稷始也"[80]。白居易录一百首元稹诗于屏风[81]。《历代名画记》有"董伯仁、展子虔、郑法士、杨子华、孙尚子、阎立本、吴道玄屏风一片，值金二万，次者售一万五千"之说[82]。

隋唐时还有一种屏具，是佛像的背屏，见于敦煌405窟隋代《佛说法图》壁画，在龙门石窟惠简洞西壁弥勒佛和山西博物院藏天宝四年石弥勒佛像背侧亦见雕有这种背屏。屏在高度近佛像头部处以横桄界为上下两部分，横桄两头饰龙首（惠简洞弥勒像背屏未见），上部为连弧形组成的半圆状，下部则以横桄格成数格，中有圆光，饰有图案。这种背屏应该是从佛像背后的木架搭织物而成的障子发展而来，前述石雕所见二例，框架雕刻表达清晰，已经转化为固定的屏，是宋元时佛像后背屏或椅背造型的源头。肃南唐墓出土有木制残件，为2厘米见方的帐子攒成的方格，上有连弧形横木，其上有镶装的铜鎏金龙头和桃形饰，由于挖掘时已有破坏，据乡民回忆认定为床的残件，从造型来看应是家具（或车舆）后侧屏状造型，与此处佛像背屏相近，是此类屏的实证[83]。

五代南唐画家王齐翰《勘书图》（南京大学图书馆藏）中出现了一种新形式的大型座屏〔图11〕，自魏晋时的三面围屏发展而来，但已从"冂"形发展为"⌒"形，自带屏墩，独立于床榻之外。画中屏的尺寸巨大，应为木胎糊绢而成，皆有甚宽的边框，横竖框格角相接。中扇扁阔，横向构图，边扇竖向构图，绘精致的山水图画。中扇和两侧扇各以梭形四花瓣状合页相连，一排四个，显然此屏可以向前任意折叠。屏墩为卷云状，侧面下方雕饰壶门，上方中间开槽，屏板插入槽内固定。传五代周文矩《重屏会棋图》（故宫博物院藏）中屏上所绘"重屏"，亦是类似样式，屏墩则为趴伏的狮子状〔图12〕。

《重屏会棋图》前景的榻后施"一"字形大座屏。座屏是自周时黼依发展而来，隋唐少见身影，此时又开始流行起来。榻后施大座屏的摆放方式或是继承了唐代榻后施曲屏的方式。这种陈设自此而始[84]，盛于宋，明清渐衰。就屏的结构和造型看，与西汉马王堆墓出土彩绘座屏几无大的差别。

孟蜀高祖孟知祥作嬣宫，"以画屏七十张，关百纽而斗之，用为寝所"[85]。则是以

11 | 五代 王齐翰《勘书图》卷
南京大学图书馆藏

12 | (传) 五代 周文矩《重屏会棋图》卷
故宫博物院藏

多扇围屏为室内装饰，将屏转化为碧纱橱或隔扇的做法。这种做法后世依然沿用，明代闵齐伋绘刻《西厢记》彩图第十三幅中出现一架围绕架子床的围屏，即是此类。

至迟五代时，屏风已经当门而立，如同古时"树塞门"的做法。南唐李煜在碧落宫置"八尺琉璃屏，画夷光独立图"，以至于冯延巳误认为是"宫娥着青红锦袍当门而立，故不敢径进"[86]。就是在门口安设座屏。

五 宋、辽、金、元时期的屏具

宋代是中国高型家具即垂足而坐体系成熟期，低型家具基本退出历史舞台。凡明清所见家具门类，几乎都已具备，较之唐五代家具造型的质朴单一，呈现井喷般发展，以结构、造型为装饰的手法也日益增多。此时家具相关的木器、漆器加工技术已经较为发达，可以制作出精细秀丽的家具。屏具而言，已经基本形成体系，结构和造型基本成熟，且新出现了枕屏、砚屏等中小型屏具。

辽、金政权存在时间与宋相近。辽略早，兼文化艺术发展较宋缓慢，家具较为质朴，仍然延续唐风；金较晚，家具风格与宋相近而稍粗犷。本文不刻意强调其间的差别，以宋为代表，来叙述当时的屏具状况。

元代政权统治时间不足百年，然于中国家具而言却是很重要的一个转折，奔放自由风格影响中国家具一直持续到明代中期，屏具相对而言变化较少，基本继承宋、辽、金样式。

1. 落地大座屏

宋时落地大座屏分"一"字形座屏（如无强调，下文中的座屏皆指此）和"⌒"形三扇式座屏两种，室内室外皆有。

座屏的造型和结构，在宋时有了大的发展，主要有两个方面〔图13〕：

其一，出现了屏座，即两侧屏墩以横枨连为一体，有的上下设双横枨，镶装绦环板，枨下装牙板。屏具中独有的披水牙板已开始出现。所谓披水牙板是一种在屏墩间的枨下安装的一种侧面为"八"字形的两个斜牙板，如同墙头斜面砌砖的披水而得名。

其二，在屏墩和立柱（或屏扇边框）之间增设斜枨或站牙，两边抵住屏心边框，提供三角形支撑。斜枨的出现比站牙要早些，多为连弧状，如北宋张择端《清明上河图》中有数件座屏就是采用斜枨。站牙多窄而高，形如蕉叶，当出现在北宋末期或南宋早期，南宋时期已经非常普遍，或因站牙的结构更加科学，造型更加美观，比斜枨的应用更广。在传世的宋元风格屏具或衣架（衣架是家具中唯一与屏具造型相类，甚至混淆的品类）中，也见到站牙和连弧状斜枨同施于一处的做法，属于过渡期的孑遗。

13 | 宋 马和之《女孝经》卷局部
台北故宫博物院藏

14 | 宋 马和之《女孝经》卷局部
台北故宫博物院藏

　　这两个方面的发展，标志着座屏的造型和结构在宋代已经成熟，其后元、明、清各代的发展，未出其窠臼。

　　此外，以镶嵌绦环板的方式分割屏扇，是宋代座屏新出现的一种样式〔图14〕。这种样式的屏框为复框式，即在外框之内又设一圈子框，外框和子框间联以短枨，之间的空档镶嵌绦环板，素朴者任其光素或仅雕饰如意云头、委角开光等简单图案，繁杂者有人物、鸟兽、花草等。这种座屏如马和之《女孝经》（台北故宫博物院藏）、《五山十刹图》[87]所见。此类座屏的屏心位于座屏的中上部，只占较少的面积，上方两个转角大多不再方直，出现委角造型，有的甚至将上方横枨做成连弧状，中间高而两侧低。这种座屏的绦环板为雕刻、漆饰等装饰手法提供了展示的空间，兼以屏框、站牙、屏墩、屏下牙板的形式变化，都可使屏具有丰富的装饰效果。虽然屏心是座屏毫无疑问的核心，但就装饰性而言，屏框、屏座大有和屏心平分秋色的趋势（甚至已经超过屏心）。

　　抱鼓式屏墩亦开始出现在宋时的座屏上，其样式可以上溯至唐永泰公主墓出土三彩笔架[88]，即在屏墩两侧各雕成一个竖立的扁鼓或球状，下方横木如抱，故名抱鼓。这种抱鼓式造型亦见于建筑台阶和门墩处。虽然抱鼓式屏墩渐成最具代表性的样式之一，但屏墩样式依然丰富多彩，延续自战汉以来的传统，将屏墩做成各式动物者时而见于此期。此外，还有一种装饰卷草纹的屏墩[89]。

　　宋时大座屏实例罕见。金代阎德源墓出土两件，杨木制，通高1.16米、全长2.32

米，由云头屏墩、长方屏框和方格架（即屏心木龙骨）组成[90]。同墓出土有成组的木床、木桌、木案、木几、木椅、木盆架、木巾架和木座屏模型，搁置在木供桌之上，其中木座屏据考古报告所载通高 28.8 厘米、长 25.7 厘米、宽 19 厘米，正面为大理石画屏[91]。此屏出土时位于木椅（高 20.5 厘米）后，或即为椅后所用大座屏模型。北宋元德李皇后陵地宫出土木雕器座四件，其造型即为抱鼓状，竖向上方有三个榫眼，应为容纳立柱和斜枨（或站牙所用）。此器座尺寸未知，因出土时配套的构件散架、损坏，已不可复原，但无论其为衣架还是座屏，抑或其他家具的构件，都堪可作为抱鼓式屏墩样式实例来考察[92]。

宋时绘画、壁画中的大座屏形象，已是司空见惯，如北宋张择端《清明上河图》（故宫博物院藏）、宋佚名《羲之写照图》（台北故宫博物院藏）、南宋马和之《女孝经》（台北故宫博物院藏）、传顾闳中《韩熙载夜宴图》（故宫博物院藏）、南宋佚名《中兴瑞应图》（天津博物馆藏）等，都能见到应用。此时壁画中的座屏以辽、金、元墓室壁画所见最为丰富，与前列绘画中趋同的风格不同，或古朴，或新奇，呈现出丰富多变的面貌，如河南登封市黑山沟村北宋李守贵墓壁画中的书法座屏[93]、北京八宝山辽代韩佚墓的山水座屏[94]、内蒙古敖汉旗南塔子乡城兴太村下湾子 1 号辽墓壁画中的八哥图座屏等[95]。

座屏屏心以山水画最为常见，意境旷幽，可增空间之深广，花鸟、书法也是常见题材，还出现了特殊的水纹题材，如白沙宋墓《开芳宴》中墓主夫妇像、宋佚名《妆靓仕女图》（美国波士顿美术馆藏）中所见，除了装饰外还有压火寓意[96]。

"⌒"形三扇式座屏在宋时仍在应用，如宋刘松年《罗汉图》所见〔图 15〕。太原晋祠圣母殿宝座后施有座屏，"⌒"形，屏上有帽，为连弧形轮廓，中高旁低，形若三山，每扇上方横枨皆挑出凤首，屏心为复框式，上绘水纹、太阳，屏中腰设绦环板、裙板，这是难得的北宋三扇式座屏实例。《五山十刹图》中的"灵隐寺屏风样"，亦是如此结构，只是其总高达近 6 米、宽 4 米余，是罕见的巨型座屏[97]。宋至元时期的佛道画、墓室所见座屏，还有一种装饰极为华丽者，绦环板多装饰繁华至极的花纹，龙、凤、麒麟、狮子、鲛人、童子及各式瑞兽与花草等图案，纷繁杂陈，若是实物，应该还有彩绘、贴金等髹饰。制者唯恐手不巧不细，唯恐心不诚不坚，竭尽心思与能力，施诸般本领于其上。此外，四川泸县一代宋墓壁龛上常见一种三扇式座屏，前设交椅，意指墓主之位，如奇峰镇一号墓、二号墓等[98]，皆是如此，应是宋时当地流行的一种组合。

南宋时出现了"山字屏"的说法，如陆游诗中的"纸屏山字样，布被隶书铭"（《暖阁》）和"水纹藤坐榻，山字素屏风"（《初夏幽居杂赋》）等[99]。山字屏可能即指如太原圣母殿屏风这种中高旁低的样式，明清时期"⌒"形三扇式、五扇式座屏，皆习惯称为山字屏。

15 右页图
宋 刘松年《罗汉图》轴
台北故宫博物院藏

五代时屏前设榻的流行组合，至此时改为屏前设椅，概与此时椅子大肆流行有关。当然，此时屏具应用自由而广泛，前设榻、桌、杌凳等者，皆有所见。建筑内当门或者正后方设座屏已成定式，前述宋代座屏图例中均有所见，北宋时李诫所编官方建筑规范《营造法式》中已出现"照壁屏风骨"[100]。此外还有"截间屏风骨"，是以屏来间隔室内的。建于清乾隆时的苏州西山东村敬修堂中，尚保留有座屏隔断室内的做法，是甚为难得的"截间屏风骨"实例。

唐时六曲屏风是重要的书画载体，宋时座屏则是重要的书画载体。据李溪的统计，从960年建国至靖康之变，大宋大内及东京主要管制机构展示的屏风（基本为座屏）就超过30组，大内宫廷基本都有陈设屏风的记录。画屏的题材，有龙水御屏，还有祥瑞、翎毛、花竹、动物、湖石、山水等。其中最为突出的是宋神宗时郭熙的山水，其他诸如僧传古、僧继肇、钟隐、田锡、陶裔、李隐、荀信、燕肃、蔡襄、易元吉、崔白、艾宣、丁贶、葛守昌、李宗臣、贾祥、米芾、王安中等，都有画或书写屏风的记载[101]。张择端《清明上河图》中书法座屏随处可见，甚至还有两块揭自座屏的书法屏心，被随意地拿来苫盖运货的车辆。

2. 床榻上的枕屏

枕屏自宋时开始大量使用，概与宋人在室外或亭台处使用坐榻有一定关系，搁置床旁，可以遮风挡寒，为卫生之器，亦可绘饰图案，以供卧游。陈淳《枕屏铭》："枕之为义，以为安息，夜宁厥躬，育神定魄。屏之为义，以捍其风，无俾外人，以间于中……"[102]枕屏上的图画以平远山水为主，也有花鸟、人物，亦有一时兴起之题字。潘大临有《题陈德秀画〈四季枕屏图〉五首》[103]。"（邵康节）尝过士友家，昼卧，见其枕屏，画小儿迷藏，题诗其上云：'遂令高卧人，欹枕看儿戏。'"[104]枕屏上画小儿，或与其施于床的特殊位置有关，陈造《题龚养正孩儿枕屏二首》："小屏光洁雪难如，写出嬉然同队鱼。索句拥书皆薄相，笑啼嗔喜莫怜渠。""眼中何止舞雩童，况是君家积庆重。梦里送来烦孔释，要令门户继荀龙。"注曰："养正方苦少子。"[105]明确表达了枕屏画童子是求子嗣，这概是一种绘有"婴戏图"的枕屏。

枕屏的样式分座屏式和围屏式。前者中型尺寸，搁置床头枕后，最是名副其实者，后者自魏晋时的三面围屏发展而来。座屏式枕屏形象，可见于宋佚名《荷亭婴戏图》（美国波士顿美术馆藏）、宋佚名《半闲秋兴图》（台北故宫博物院藏）、宋赵伯骕《风檐展卷图》（图16）等，造型和用法都大同小异，独幅横屏式，有抱鼓、地栿、站牙，边框较窄，镶装屏心，宽度与榻相近，如果从画面比例看，枕屏宽约70～120厘米、高约50～100厘米，施于床的一端。本书收录的黑漆螺钿孔子观欹器图座屏，就有可能是一件难得的明代座屏式枕屏实例。

围屏式枕屏或许宋以前已使用，但无枕屏之名，白居易"就日移轻榻，遮风展小屏"[106]，即是此类。苏辙《画枕屏》："绳床竹簟曲屏风，野水遥山雾雨蒙。长有滩头钓鱼叟，伴人闲卧寂寥中。"[107] 是一件绘有云雾远山、垂钓渔翁的曲屏。朱熹《家礼》中记冬至祭祀用的屏风"如枕屏之制，足以围席三面"[108]，也是指三面围合的枕屏。传顾闳中《韩熙载夜宴图》中出现枕屏两种四件，一种是在床四角设柱，柱间三面连以横栿，加绦环板，余中间为屏心，上有图画；另一种

16 （传）南宋 赵伯骕《风檐展卷图》页局部
台北故宫博物院藏

17 （传）宋 苏汉臣《婴戏图》轴局部
台北故宫博物院藏

18 明晚期 黄花梨天马寿字纹靠背椅
上海博物馆藏

则是3块独幅屏扇，围于床上。围屏式枕屏与床榻的组合，后来发展为榫卯方式相合，形成了明清时期的罗汉床。传苏汉臣《婴戏图》[109]（台北故宫博物院藏），绘庭院中一张罗汉床，为三屏式围子，其中侧围子一旁还有座屏屏墩上常见的抱鼓，正是枕屏遗意〔图17〕。中国国家博物馆曾展出一件明清之际的黄花梨镶大理石罗汉床，侧围子前亦设抱鼓，屏下有地栿，显然亦是移置自枕屏[110]。宋人《维摩诘图》（台北故宫博物院藏）中出现的罗汉床则是一种复框式座屏与须弥座式床座结合的五屏式罗汉床。同样，明清时期的架子床中亦能看到床围子移植座屏的做法。

宋时还有一种施于禅座上的座屏，诸如山西高平开化寺壁画所见，这种屏与禅座的结合方式，发展成为一种新型的屏背式座椅，诸如上海博物馆藏明晚期黄花梨天马寿字纹靠背椅〔图18〕，其靠背上屏座的站牙、抱鼓、地栿俱备，就是对这种组合形象的继承。

3. 砚屏

砚屏是宋时出现的又一重要屏具，与文人好砚、好石的风尚不无关系。砚屏的应用开启了置屏于几案的新用法，今所见传世古代座屏实物，亦以几案陈设的中小型屏具为主。南宋赵希鹄《洞天清录》"砚屏辨"："古无砚屏。或铭砚，多镌于砚之底与侧，自东坡、山谷始作砚屏，既勒铭于砚，又刻于屏，以表而出之。山谷有乌石砚石屏，今在婺州义乌一士夫家……"[111]赵希鹄还罗列了宣和玉屏、永州石屏、蜀中松林石屏等。蜀中松林石"解开自然有小松,形或三五十株行列成径,描画所不及,又松止高二寸,正堪作砚屏。屏之式,止须连腔脚高尺一二寸许,阔尺五六寸许,方与盖小砚相称,若高大非所宜,其腔宜用黑漆或乌木,不宜用钿花、犀牛之类"[112]。明确提出砚屏屏框(即屏腔)高约4~8厘米、宽约16~20厘米[113],与砚的尺度相仿。此外,"取名画极低小者嵌屏腔亦佳,但难得耳"[114]。可见砚屏已不只镶嵌石板,还可镶画或背后裱画。宋时砚屏图像和实例少见,张择端《清明上河图》(故宫博物院藏)近末段"久住王员外家"右侧楼上,绘一个背靠书法座屏的读书人,其书案上就摆着一个正对他的小砚屏。南宋李嵩《四迷图·酗酒》中砚旁绘有小座屏,尺余见方,复框式,屏心素而无纹,下有一排三个绦环板,是较具体的宋元座屏形象[115]。至若宋佚名《道子墨宝:地狱变相图》(美国克利夫兰艺术博物馆藏)中,除了丰富的落地大座屏外,在座屏后的桌案上,就出现了三个样式各异的中小型案上座屏,站牙、抱鼓墩等构件清晰可见。明谢环《杏园雅集图》图中砚屏使用方式,亦可作为宋时砚屏的注脚〔图19〕。

19 | 明 谢环《杏园雅集图》卷局部
美国大都会艺术博物馆藏

宋时镶石砚屏负盛名者颇多，首推欧阳修的月石砚屏。欧阳修作《月石砚屏歌》，歌序描述"一石中有月形，石色紫而月白，月中有树森森然，其文黑而枝叶老劲……月满西旁微有不满处，正如十三四时，其树横生，一枝外出"[116]。歌中对石屏赞誉备至，"月从海底来，行上天东南。正当天中时，下照千丈潭。潭心无风月不动，倒影射入紫石岩"[117]。对石上图案的认知充满了文人浪漫的想象。苏舜钦作《永叔月石砚屏歌》以和[118]。梅尧臣则作《读月石诗》，对二人关于月石砚屏的诗论提出异议。此后苏轼、苏辙都有关于月石砚屏的诗歌，苏轼却认为三人"无事自作雪羽争"。苏轼亦有月石林砚屏、雪林砚屏等，而宋人还有多首砚屏诗歌传世，可见砚屏应用之盛。月石砚屏引发的是当时人关于石屏在哲学、美学乃至文学方面的认知，暂不在本文谈论范围。就砚屏本身来看，石纹具自然山川之美，如画如幻，"床头复一月，下有风林横"，也许这才是砚屏盛行于士大夫之间的原因。

宋时砚屏用材以虢州石最常见。《云林石谱》载："虢州荥阳县石，产土中，或在高山，其质甚软，无声。一种色深紫，中有白石，如圆月，或如龟蟾吐气、白雪之状，两两相对。土人就石段揭取，用药点化镌治而成，间有天生如圆月形者极少得之……又有一种色黄白，中有石纹如山峰罗列，远近涧壑，亦是成片。修治镌削，度其巧，辄乃成物像，以手砻之，石面高低。多作砚屏，置几案间，全如图画。询之土人，石因积水浸泽，遂多斑斓。"[119]虢州石出河南省灵宝市、卢氏县一带（即古虢州所在），自唐始作砚。欧阳修、苏轼等人的月石砚屏是一种紫地白斑者，今所见明清座屏中有镶绿石者，有观点认为即虢州石，石纹多作黄绿色，其上山水、人物、鸟兽之象，多为磨制而成[120]。《清异录》："京城北医者孙氏有木颊小石屏，石色赤绿，上有正白如蒙头坐僧，颇类真。京人相沿号玉罗汉屏孙家。"[121]京城医家小屏成了家喻户晓的名品，可见砚屏在世俗间应用之广。

4. 围屏

多扇围屏在宋、辽、金、元时仍在使用，但远不如唐时六曲屏风之盛，或因功能与围屏式枕屏相近，后者盛行，取而代之。北京市石景山八角村金墓壁画《侍寝图》中出现了一种新式的绕床围屏，屏设有望柱头的四柱，围于床榻四角，形成侧三面背五面共计11屏的围合状（其中背屏中间一扇较宽，右侧转角处两屏各画一半，或是图画误差，或是形制有特殊处，尚待研究），每扇屏上方连弧状边框，上方尺许又有三个连弧状弯枨相连，甚为奇特[122]。山西朔州市政府工地辽墓壁画，也见有类似造型的围屏[123]。至明清时期，围屏已经发展成为成熟的独立体系，无论造型还是功用，都与座屏有了较大区别，故本文不再叙述。

20 | 明 佚名《上元灯彩图》卷局部
台北观想艺术中心藏

六 明、清时期的屏具

明代中期以前的家具，尚延续宋元样式，屏具亦然，如朱檀墓出土木座屏模型，尚是宋元风格。明中期以后，小商品经济发达，以黄花梨为首的硬木制家具开始流行，中国家具中精致细腻的硬木家具与自由奔放的软木家具相互争辉。中国高型家具发展已至高峰，门类齐备、结构成熟、装饰多样、工艺繁杂。书画艺术也不再以屏具作为主要的展示方法，但是各种工艺、材质尽可施于屏心，故而对其他传统工艺的综合性，是任何家具门类都难以企及的。此时的座屏随处可用，市肆也有各式屏出售〔图20〕。

1. 明清时期座屏的材质

以材质而言，座屏的框、座多以木为主，有黄花梨、紫檀、乌木、黄杨木、檀香木、铁梨木、鹨鹅木、柞榛木、柞木、榉木、榆木、柏木、核桃木、杉木、松木等，还有少见的瓷、石、铜、铁、珐琅、玉等，漆饰又有黑漆、红漆描金、彩画、戗金、雕漆、螺钿、百宝嵌、嵌银丝等。屏心的内容，更是丰富，除了前述屏框上采用的工艺皆可外，还有绿石、大理石、竹叶玛瑙石、祁阳石、松花石、玉、青金石、玛瑙、瓷、珐琅、木雕、竹簧、竹雕、书画、织绣、烫画、铁画、百宝嵌、画玻璃、刻玻璃、料丝、玻璃镜、嵌牙、点翠、沉香、檀香等。

屏心镶石以绿石、大理石、祁阳石为大宗。绿石类的虢州石前已述及，宋人取作砚屏。所见明清座屏实例采用绿石板者，多将之依纹路磨制为各式图案，以山水间人物、鸟兽出没者多见，有的颇合画意，有的则天真烂漫，本书即收录多例。

大理石以产云南大理得名[124]，黑质白章，纹路如山峦、云气，和米家山水意境最合，是制作屏风、桌面的上好材料，明清时盛行。文震亨认为"屏风之制最古，以大理石镶下座精细者为贵"[125]。《金瓶梅》中记应伯爵劝西门庆当的一架大螺钿大理石屏风和两架铜锣铜鼓连铛儿，值30两银，大理石屏风"三尺阔、五尺高，可桌放的螺钿描金大理石屏风，端的是一样黑白分明"。应伯爵夸赞"恰相好似蹲着个镇宅狮子一般"。"西门庆把屏风扶抹干净，安在大厅正面，左右看视，金碧彩霞交辉"[126]。大理石开采，目前所知资料，最早始于明景泰年间，甚为贵重，曾作为珍奇异宝进献宫廷[127]。清代亦有以广东地区产大理石镶嵌座屏者，纹路青灰，稍逊。所见传世大理石屏，石纹甚为丰富，仅色泽而言，除了黑色石纹外，尚有灰、青、绿、黄等色，或多色交叠者，各擅胜场，往往出人意表，令人拍案。

祁阳石产湖南永州，故又名"永石"，质地细腻，可作砚材，但不甚坚，其石多作紫红色，夹杂有绿、白、黄等层，匠师留不同层为各式颜色，巧雕为图画，精致艳丽。所见镶祁阳石座屏，明代者古朴，颜色较单一，清代者妍美，花色纷繁，若赋彩而成。前述绿石亦有一些被认为是祁阳石者。

料丝灯屏，明初即有。郎瑛《七修类稿》载："料丝灯出于滇南，以金齿卫者胜也，用玛瑙、紫石英诸药捣为屑，煮腐如粉，然必市北方天花菜点之方凝，而后缫之为丝，织如绢状，上绘人物、山水，极晶莹可爱，亦珍贵，盖以煮料成丝，故谓之料丝。"[128]《韵石斋笔谈》："丝灯之制，始于云南。宏（弘）治间，邑人潘凤，号梧山，善丹青，有巧思，随杨文襄公至滇中，见料丝灯，悦之，归而炼石成丝，如式仿制，于是丹阳丝灯，达于海内。"[129] 陆深作《宝丝灯屏歌》，称料丝"云腴石髓蟠青葱，酒黄鸦紫祖母绿。复有无价桃花红，远人不贡车与服。持此射利如射鹄，遂令长安富豪家。下视隋珠贱和玉，琐屑璀璨星宿同"[130]。可见料丝在当时之流行。料丝近似今之玻璃丝，因其透明，放置灯旁，既不遮挡亮光，又可蔽风，自然是制灯屏的上好材料。故宫博物院存有清晚期楸木框料丝四扇屏[131]，查其样式，是用细如发丝的料丝斜向密排前后两侧，方向刚好交叉，但并不相接，中间夹层为剪贴而成的花鸟图案，有灯则透光，图画清晰可见。

瓷屏的制作可能宋时已有，但暂未见可靠的实例[132]。明正德时，景德镇制作一种青花瓷屏心，中间多为圆光内绘阿拉伯文，周匝绘卷草纹绦环板，两侧有凸起的榫舌，可以插入屏座。还有一种青瓷或青花制砚屏，前方多作高台，上有孔可插笔，屏背多饰魁星点斗，是读书人喜用的一种小座屏。自明中期以后，瓷屏和镶嵌瓷板的屏具渐多，各式釉色皆有。清代还制作一种仿木纹釉的瓷座屏，屏心描绘图案，屏框彩绘为木纹。

2. 明清时座屏的造型

明清时常规的座屏由两大部分组成，即屏心所在的屏扇和下面稳定用的屏座。屏扇包括屏心和屏框以及屏框上的绦环板；屏座则由屏墩、站牙（或斜枨）、披水牙板、绦环板等组成〔图21〕。插屏属座屏的一种，屏座与屏扇为拔插式可分离，凡座屏有的造型、装饰等皆有，但插屏需在屏墩上另设立柱，镶装屏扇。立柱造型及立柱与屏扇间交接方式的变化，则是插屏独有的。

屏扇

明清时座屏上屏扇的样式已甚富变化。最简单者为素边框者，有宽窄、薄厚之变化，形成迥异的意趣。若是边框上再施以打洼、委角、素混面、剑脊棱、皮条线、阳线、两炷香线、阴线、回纹等装饰，或简或繁，更具变化。自宋已有的复框式屏扇此时应用更广，基本样式是在屏心边框外又设外框，之间余寸许空间，以短枨隔开为横竖空档，镶以绦环板。外框上转角处多做成委角，横枨与委角一木挖成，如同烟袋锅榫般与竖枨相接。框的线脚，以剑脊棱、皮条线相结合者最为典型，有些考究者双面可观，正面边框线脚亦各不同。复框式屏扇始于宋，盛行至明末清初，嗣后渐少。

绦环板更适装饰，各式图案无不适宜，简朴者只做剑环式、海棠式、如意云头式开光，或浮雕，或透雕，复杂者则透雕、高浮雕兼施，花鸟人物无所不有。明式黄花梨座屏，以饰螭龙纹者最为多见。此外，亦有一些镶装石板，代替木制绦环板者。

屏扇上屏框与屏心的结合，基本都采用传统的攒框结构，四框做槽口，屏心以榫舌纳入边框，也偶见一些屏背加设一圈子框，设活销，可以取换屏心者。

案上座屏，多为单扇独幅，落地的座屏则有单扇、三扇、五扇甚至更多者。文雅者多设单扇，与宋时用法类似，镶装书画其上。宫廷和官府多用三扇，气势雄伟，彰显威严，图案则麒麟纹、祥云拱日纹等，帝王则用云龙纹，彰显身份。

21 | 明晚期 红漆框镶绿石座屏
庶园藏

屏墩与站牙（或斜枨）

屏墩多厚重，与站牙结合，形成一个三角结构，起到稳定屏扇的作用。

简单的屏墩直接置横木一块即可，较复杂者常见有两种：一种为拱桥式，将一块横木下方锼挖出亮脚，上方如罗锅般拱起，两边如足落地。在此基础之上，又可因亮脚的不同、足的变化，形成各种拱桥式变体的造型，甚至有的侧面观之如同小几。另一种为抱鼓式，前已述及，宋时屏具中已经广泛使用，究其来源，鼓形实若日月，下方雕云纹，有云拱日月的意向，至若其脉络，尚需更进一步整理。抱鼓墩的样式，自宋至明清，虽然大体样式没变，但比例、意趣也发生了很大的变化，宋时鼓状构件较大，在屏墩上的位置偏上，靠近屏扇竖枨，相应的鼓状构件下的部分形如台座，高高拱起，下方挖小壸门。明清时鼓状构件较小，且位置偏下而位于两端，下部较矮，两头多做成卷转环抱的造型。鼓状构件侧面，或任其平素，或多雕饰葵花瓣，如同旋涡，是最为经典的装饰手法。此外，一些异形的屏墩也出现在考究的座屏上，清代晚期流行雕刻狮子为坐墩的座屏，喜气洋洋，但格调不高。

斜枨或站牙是连接抱鼓墩与屏扇竖枨间的构件。斜枨出现较早，明清时已渐为牙板所取代，相对较少见，大型的座屏甚至会采用金属质斜枨加固。站牙出现在南宋时期，其发展与抱鼓构件的变化相呼应，宋时抱鼓大而站牙小，明清时抱鼓渐小而站牙变大。宋代站牙多为三角形带连弧形边缘，窄小，形如小蕉叶，明代常见桨腿式、壸瓶式、半石榴头式、如意云纹式等，清代则变化纷繁，只要是大体轮廓呈三角形（有的甚至三角形都不是），一切造型、装饰都可，不唯透雕、浮雕，甚至圆雕如蹲龙、戏狮、卧熊等，也出现在站牙的位置。

绦环板

复框式座屏镶装绦环板前已述及，还有一种单框屏扇在下方设绦环板者，在明晚期以后的座屏中甚为常见。其形式为在屏墩之间，设双横枨，之间镶装绦环板，既有通长者，也有以矮老界成两档或者三档者，可以自由发挥，施加各种装饰。这种下方设绦环板的做法，多出现在插屏中，因为插屏的底座和屏扇可以分离，双横枨加绦环板的样式可使底座更加牢固。在明代或清早期版画中，常看到一种下方镶嵌很高的云头纹等样式的绦环板，进而将整个屏心抬高至上半部的大座屏，但传世家具中少见。

牙板或披水牙板

屏座的下方横枨下多设牙板或披水牙板。牙板多为竖向单牙板，以刀牙板或壸门牙板最为常见，偶见一些特殊样式，相对还是单一，多用在一些风格简朴的家具之上。个别座屏上采用双层牙板的做法，即前后各装牙板，中间留空槽。

披水牙板是座屏中经典的构件，形如"八"字前后撇开，给人以稳固协调的观感。其样式亦是凡桌案牙板能有者，几乎皆有，其中以刀牙板、壸门式最为多见。此外还有将披水牙板表面做成弧面者。

立柱

立柱是插屏特有的结构，即在屏墩上竖立一段帐子，内侧开槽，容纳屏扇。立柱的造型有的和屏扇造型合为一体，内里以榫卯咬合，有的则为明显的柱状，上方做成宝瓶、石榴头、仰俯莲、蹲狮等造型，和栏杆上的望柱相若。

地栿

有些考究的座屏，屏墩下另附有地栿，多呈"亚"字形，横向还衬托披水牙板。一般在大型落地座屏中常用到地栿，年久腐蚀严重更换即可。此外还有在两个屏墩下单设一道者，与须弥座底端的圭角相近。

3. 明清座屏的功用

陈设于地的大座屏，仍是黼依遗意。"一"字形座屏明清版画中甚为常见，多立于厅堂正后方，前设椅、榻、桌等，这种大座屏保存不易，至今少见实物流传，所见精彩的实例当属美国加州古典家具博物馆所藏清代黄花梨镶大理石螭龙纹座屏，高214厘米、宽181厘米、深105厘米，绦环板透雕螭龙纹，玲珑剔透，镶装纹路优美的大理石屏心，气势撼人，是少有的传世大型座屏[133]。"冖"字形座屏多设在宫廷或官府、庙宇的大殿，多为三扇或五扇的山字形座屏，亦有多扇的座围屏。屏上方多有屏帽，屏座多为须弥座式，更加稳重。屏心在宫廷以龙凤等皇家特征明显者为多见，官家则以麒麟或海上红日为主，也有山水、花鸟、人物等，多有祝寿、祈福等吉祥寓意。明代以漆饰为主，清代太和殿等重要场合为金漆，其他殿宇多设硬木座屏，屏心材质丰富。传世的明清大座屏中还有一种三扇式，中间形如座屏，旁边两扇形如门扇，与之以铰链相合，既可打开为山字形座屏，又可折叠如为"一"字形座屏。这类保存不易，所见者寥寥，案头陈设的小座屏亦有相同样式者。造办处档案记载清雍正时期还制作一种大型座屏，正面做成多宝格式，可以展示古董珍玩，是兼具架具的功能，今故宫博物院藏有与之制作时间和造型相当的紫檀黑漆描金多宝格式座屏[134]。

陈设于室外的座屏，见于紫禁城永寿宫和景仁宫，当门处各设有一件尺寸相同的汉白玉框、大理石心的座屏〔图22〕，大理石纹路或如飞龙，或如山峦，是罕见的明代室外大座屏孑遗。紫禁城养心殿门外当门陈设一件制作于乾隆时期的铜座屏，饰以云龙，镶以玉璧，既是皇权之象征，又是皇帝工作时所见以自省者。

中型和小型座屏陈设于几案，前者多设于大案上，平时不常移动，后者书斋案头、厅堂边几，皆可随意而设，护砚为砚屏，护灯为灯屏。清代喜在中堂条案上摆放座屏，寓意"平安"。

笔架和座屏结合的方式前已述及瓷质者，木质者亦有数件。明代晚期朱守城墓

出土有一件极为精彩的紫檀镶大理石座屏，高20厘米、长17厘米、宽8厘米，不设屏墩，设较宽的托泥板，前方做成一件倚靠座屏的条桌形，面上有孔，可以插笔四管，屏框简素，桨腿式站牙，大理石黑白分明，纹路如画，清新跳脱〔图23〕。

明清家具中还常见座屏与其他门类相结合者，诸如座屏进深加宽，前设两门或插门，内为箱体，甚至设抽屉，可以储藏画卷、宝物，今故宫博物院所藏兰亭八柱卷即收藏于一件有暗抽屉的紫檀座屏之中。也有座屏内装锡胆和玻璃，可作鱼缸者[135]。有一种条案的腿足也做成座屏式，屏墩、披水牙板等座屏的经典造型皆备。清晚期还出现一种屏背椅，多见于江南地区，是在机凳上设屏板状结构，前方多设雕饰灵芝纹或夔龙纹的站牙，屏心或镶大理石，或阴刻花鸟、山水图案，虽有前述开化寺禅座后设屏的遗意，但整体造型的格调不高。至于宝座，扶手和后背则多移植座屏的样式。山西地区有一种火炕配套用的挡火石，多雕作座屏状，造型拙朴，至迟明代已有。至若以座屏造型制成几架者，更是举不胜举，皆因其座墩与站牙形成的三角形结构，可以将平面物品稳固竖立。

座屏的功用，除了屏蔽、遮挡和陈设外，还可用来题记书写，作记事之用。《水浒传》记柴进簪花入内廷，"到一个偏殿，牌上金书'睿思殿'三字，此是官家看书之处。侧首开着一扇朱红槅子，柴进闪身入去看时，见正面铺着御座，两边几案上放着文房四宝：象管笔、花笺、龙墨、端溪砚，书架上尽是群书，各插着牙签，勿知其数。正面屏风上堆青叠绿，画着山河社稷混一之图。转过屏风后面，但见素白屏风上，御书四大寇姓名。写着道：山东宋江，淮西王庆，河北田虎，江南方腊"[136]。《国学礼乐图》中记"祝版"[137]，也是座屏样式，"漆木为之，外黝，中质以纸书祝文面粘之，祭毕则捧祝文以焚"。民间座屏多有上书"天地君亲师"等神位者，用以祭祀。

屏具是独具东方魅力的家具品类之一，历史悠久，绵延不绝，今日所知，既有一脉相承，千余年来几无变化者，又有精彩纷呈，发展出各种形式者。就本文内容而言，以座屏为主，略记其纲目，然其广博、深邃，非一文可尽，况乎水平有限，只鳞片爪，挂一漏万，权作抛砖引玉之用，希望将来有关屏具的研究越来越深入。

23 | **明晚期 紫檀镶大理石座屏**
明朱守城墓出土

22 | 左页图
明 汉白玉镶大理石座屏
紫禁城景仁宫

注释：

〔1〕（汉）郑玄注：《纂图互注礼记》卷八，四部丛刊景宋本。

〔2〕（汉）司马迁：《史记》卷七十五"孟尝君列传第十五"，清乾隆武英殿刻本。

〔3〕湖南省博物馆、中国科学院考古研究所：《长沙马王堆一号汉墓发掘报告》上册，文物出版社，1973年。

〔4〕（汉）郑玄注：《礼记》卷一"曲礼下第二"，四部丛刊景宋本。

〔5〕（汉）郑玄注：《礼记疏》"附释音礼记注疏卷第五"，清嘉庆二十年南昌府学重刊本。

〔6〕（汉）郑玄注：《周礼》卷五"春官宗伯第三"，四部丛刊明翻宋岳氏本。

〔7〕（战国）吕不韦：《吕氏春秋》第六卷"季夏纪第六"，汉高诱注，四部丛刊景明刊本。

〔8〕（汉）班固：《汉书》卷二十二"礼乐志第二"，唐颜师古注，清乾隆武英殿刻本。

〔9〕唐友波：《春成侯盉与长子盉综合研究》，《上海博物馆集刊》第八辑，2000年。

〔10〕（晋）郭璞：《尔雅》卷中"释宫第五"，第40页，四部丛刊景宋本。

〔11〕（汉）郑玄注：《周礼》卷二"天官冢宰下"，四部丛刊明翻宋岳氏本。

〔12〕唯不同的是，宋元以来，祭祀所设屏，不只代表天帝，各路仙佛、圣贤，乃至逝去的先祖、亲眷，都可。

〔13〕楚墓出土有一种瑟座，常被误作"座屏"，典型者如湖北江陵楚墓出土彩漆木雕动物纹座屏（见湖北省文化局文物工作队《湖北江陵三座楚墓出土大批重要文物》，《文物》1966年第5期），长51.4厘米、宽12厘米、高15厘米。相关研究见范常喜《楚墓出土瑟座用途与名称重探》（中国文化遗产研究院：《出土文献研究》第十五辑，第56～72页，2019年），据该文揭示，这类瑟座在战国楚地名"瑟毁（梡）"，汉代名"瑟禁"。从造型看，这种瑟座接近后世镂雕小座屏。然以战国至汉时屏具发展状况看，并无此类，后世屏具亦未发现与其有直接联系。小型座屏要至宋时才开始应用，也才开始案头陈设，但无此矮而长者。

〔14〕河北省文物研究所：《䨟墓——战国中山国国王之墓》上册，第259～270页、279页，文物出版社，1996年。

〔15〕需商榷的是：之前将之摆放方式复原为夹角居中，两屏斜置，如"八"字形，应不确。一则这种围蔽后侧和一侧的"L"形曲屏在汉代甚为流行，可证诸壁画、画像石等图像（后文述及），未见有夹角居中者。二则屏具在室内的布局，如汉李尤《屏风铭》所言"舍则潜避，用则设张，立必端直，处必廉方"，不应出现这种布局。该屏合理的摆放方式应是一侧在后（错金银牛屏座所在一侧）为背屏，一侧在左为侧屏（错金银犀屏座所在一侧，此时夹角处的错金银虎噬鹿屏座的虎首朝前），是典型"L"形曲屏。

〔16〕（汉）佚名撰，（清）孙星衍校：《燕丹子》卷下，第25页，辑入《汉魏六朝笔记大观》，上海古籍出版社，1999年。

〔17〕（汉）刘歆撰，（晋）葛洪集，王根林校点：《西京杂记》卷一"昭阳殿"，第82页，辑入《汉魏六朝笔记大观》，上海古籍出版社，1999年。

〔18〕（汉）刘歆撰，（晋）葛洪集，王根林校点：《西京杂记》卷一"飞燕昭仪赠遗之侈"，第84页，辑入《汉魏六朝笔记大观》，上海古籍出版社，1999年。

〔19〕（汉）刘歆撰，（晋）葛洪集，王根林校：《西京杂记》卷二"四宝宫"，第91页，辑入《汉魏六朝笔记大观》，上海古籍出版社，1999年。

〔20〕（汉）刘歆撰，（晋）葛洪集，王根林校：《西京杂记》卷六"文木赋"，第115页，辑入《汉魏六朝笔记大观》，上海古籍出版社，1999年。

〔21〕（前秦）王嘉撰，（梁）萧绮录，王根林校点：《拾遗记》卷五"前汉上"，第526页，辑入《汉魏六朝笔记大观》，上海古籍出版社，1999年。

〔22〕王利器校注：《盐铁论校注（定本）》（上）卷六"散不足之第二十九"，第356页，中华书局，1992年。

〔23〕（汉）班固：《汉书》卷六十六"陈万年传"，清乾隆武英殿刻本。

〔24〕（南朝宋）范晔：《后汉书》卷三十三"朱冯虞郑周列传第二十三·郑弘传"，第2030、2031页，百衲本景宋绍熙刊本。

〔25〕纪亮"为尚书令，而陟（纪亮之子）为中书令，每朝会，诏以屏风隔其座"。见（晋）陈寿撰、（南朝宋）裴松之注《三国志》卷四十八"吴书三"，第2252页，百衲本景宋绍熙刊本。

〔26〕湖南省博物馆、中国科学院考古研究所：《长沙马王堆一号汉墓发掘报告》上册，文物出版社，1973年。

〔27〕此座屏的方向多被错误地认为龙纹一侧为正面，从出土报告中可以明确看到穿璧纹一侧为正面。马王堆二号西汉墓出土类似彩绘座屏，亦是穿璧纹在前。

〔28〕湖南省博物馆、中国科学院考古研究所：《长沙马王堆一号汉墓发掘报告》上册，文物出版社，1973年。

〔29〕依次见：1.傅举有、陈松长：《马王堆汉墓文物》第67页，湖南出版社，1992年。2.武威县文管会：《甘肃省武威县旱滩坡东汉墓发现古纸》，《文物》1977年第1期。该屏尺寸、造型皆与马王堆一号墓座屏模型相近，屏心为粗纱为之（已腐朽）。3.洛阳博物馆：《洛阳涧西七里河东汉墓发掘简报》，《考古》1975年第2期。该屏为陶模型，结体与马王堆一号汉墓彩绘座屏模型相近，不同处是此屏为竖屏式（纵向大于横向），且屏墩位于底边

两端。

〔30〕汉乐浪郡墓葬出土，〔日〕梅原末治等：《朝鲜古文化综鉴》第二卷，图版二九，养德社，1948年。转引自扬之水《唐宋家具寻微》，人民美术出版社，2015年。

〔31〕徐光冀：《中国出土壁画全集·3·内蒙古》第31页，科学出版社，2012年。

〔32〕李文信：《辽阳发现的三座壁画古墓》，《文物参考资料》1955年第5期。

〔33〕中国画像石全国编辑委员会：《中国画像石全集·1·山东汉画像石》第160页，河南美术出版社，2000年。

〔34〕广州市文物管理委员会、中国社会科学院考古研究所、广东省博物馆：《西汉南越王墓》附录一《南越王墓出土屏风的复原》，文物出版社，1991年。

〔35〕（汉）班固：《前汉书》卷二十二"礼乐志第二"，唐颜师古注，清乾隆武英殿刻本。

〔36〕（汉）刘歆撰，（晋）葛洪集，王根林校：《西京杂记》卷四"梁孝王忘忧馆时豪七赋"，第103页，辑入《汉魏六朝笔记大观》，上海古籍出版社，1999年。

〔37〕孙机先生将这种"L"形屏具称为屏扆，后为扆，侧为屏，左尊而右卑，使用者自右登榻（或席）。见孙机《中国古代物质文化》四"建筑与家具"，第163页，中华书局，2015年。

〔38〕清代罗汉床中出现一种二面围屏的样式，清晚民国时海派家具中的贵妃榻亦是两面围屏，属当时的创新之举，与这种曲屏并无继承关系。

〔39〕（宋）李昉、李穆、徐铉等：《太平御览》卷七百一"服用部·屏风"，清文渊阁四库全书本。

〔40〕（汉）刘歆撰，（晋）葛洪集，王根林校：《西京杂记》卷四"梁孝王忘忧馆时豪七赋"，第103页，辑入《汉魏六朝笔记大观》，上海古籍出版社，1999年。

〔41〕（宋）李昉、李穆、徐铉等：《太平御览》卷七百五十，清文渊阁四库全书本。

〔42〕（晋）崔豹撰，王根林校点：《古今注》"杂注第七"，第247页，辑入《汉魏六朝笔记大观》，上海古籍出版社，1999年。

〔43〕（晋）陆翙：《邺中记》，钦定四库全书本。

〔44〕相关的讨论见巫鸿《重屏——中国绘画中的媒体与再现》第三章"内在世界与外在世界"，第162、163页，文丹译，黄小峰校，上海人民出版社，2017年。

〔45〕济南市博物馆：《济南市东郊发现东魏墓》，《文物》1966年第4期。

〔46〕山西省大同市博物馆、山西省文物工作委员会：《山西大同石家寨北魏司马金龙墓》，《文物》1972年第3期。

〔47〕漆屏陈设时哪一面在前尚待研究，出土时朝上一面为列女故事。

〔48〕根据出土报告描述，出土时较完整者5块，还有大小不一的9片残漆片较清楚，只足8片屏板之数，是否果足12片，尚需更多信息的披露。

〔49〕杨泓《漫话屏风》中将司马金龙墓出土的石雕柱础复原为漆屏底座，似不妥，柱础应为帐杆座。见易水《漫话屏风——家具谈往之一》，载《文物》1979年第11期。此套漆屏应附有边框后直接施于小榻使用。另，亦不排除是每面4片漆板，三面等宽且近方，直接施于席周匝，类似于传顾恺之《列女仁智图》卷所见。

〔50〕此石棺床未见详细信息，可能为石围屏榻式，后文所述图案应属石围屏之上者。见周到主编《中国画像石全集·8·石刻线画》第52～59页，河南美术出版社，2000年。

〔51〕王银田、刘俊喜：《大同智家堡北魏墓石椁壁画》，《文物》2001年第7期。

〔52〕此图画上围屏围合的方式，考虑绘画透视关系不确的因素，或许是一种四面合围，前面两扇开合门扇的围屏。

〔53〕需注意的是，目前所知，尚未发现明代以前使用架子床的资料，从结构上看显然是从床榻与帷帐的组合关系发展而来，但无法确定从《女史箴图》卷所绘围屏、帷帐的床到后世架子床之间是否有清晰的继承关系。

〔54〕张庆捷：《献给另一个世界的画作——北魏平城墓葬壁画》插图9，收入上海博物馆编《壁上观——细读山西壁画》第82～95页，北京大学出版社，2017年。

〔55〕山西省考古研究所、太原市文物考古研究所：《太原北齐徐显秀墓发掘简报》，《文物》2003年第10期。

〔56〕山东省文物考古研究所：《济南市东八里洼北朝壁画墓》，《文物》1989年第4期。

〔57〕林圣智曾对这套围屏做过仔细的整理和阐释，详见林圣智《图像与装饰——北朝墓葬的生死表象》，台北"台湾大学"出版中心，2019年。

〔58〕西安市文物保护考古所：《西安北周康业墓发掘简报》，《文物》2008年第6期。

〔59〕陕西省考古研究所：《西安北郊北周安伽墓发掘报告》，《考古与文物》2000年第6期。

〔60〕与前述《女史箴图》所见围屏、床和帷帐组合与架子床的关系相似，石围屏榻开宋时出现的罗汉床先声，但与罗汉床之间亦不能确定有直接的继承关系。

〔61〕天水市博物馆：《天水市发现隋唐屏风石棺床墓》，《考古》1992年第1期。

〔62〕（唐）李贺：《李贺诗歌集》"诗歌篇第二"《屏风曲》，四部丛刊景金刊本。

〔63〕（唐）李商隐：《李义山诗集》卷之六"五言绝句·屏风"，四部丛刊景明嘉靖本。

〔64〕如第335窟、220窟、234窟初唐壁画，第103窟、172窟、194窟盛唐壁画，第195窟中唐壁画，第9窟、12窟（团窠）晚唐壁画（图14）等。

〔65〕如盛唐217窟壁画《妙庄严王本事品》、23窟壁画《化城喻品》、172窟壁画《未生怨》、148窟壁画《药师净土变》，中唐44窟壁画《涅槃经变之摩耶夫人奔丧》，晚唐156窟佛像等。

〔66〕据杨泓先生的综合分析，有六曲屏风的墓葬壁画以山西太原唐墓中出现最早，约在武则天当政晚期，其次在长安附近流行，以树下男像人物为最多，似有故事情节，至会昌年间，以翎毛、云鹤为装饰者渐多。见杨泓《逝去的风韵——杨泓谈文物》"'屏风周昉画纤腰'——漫话唐代六曲屏风"，第40、41页，中华书局，2007年。

〔67〕徐光冀：《中国出土壁画全集·7·陕西》下册，第393～398页，科学出版社，2012年。

〔68〕井增利、王小蒙：《富平县新发现的唐墓壁画》，《考古与文物》1997年第4期。

〔69〕徐光冀：《中国出土壁画全集·9·甘肃·宁夏·新疆》第214页，科学出版社，2012年。

〔70〕徐光冀：《中国出土壁画全集·9·甘肃·宁夏·新疆》第216页，科学出版社，2012年。

〔71〕山西省文物管理委员会：《太原南郊金胜村唐墓》，《考古》1959年第9期。

〔72〕山西省文物管理委员会：《太原市金胜村第六号唐代壁画墓》，《文物》1959年第8期。

〔73〕山西考古研究所：《太原市南郊唐代壁画墓清理简报》，《文物》1988年第12期。

〔74〕马晓玲：《北朝至隋唐时期墓室屏风式壁画的初步研究》，西北大学2009年硕士毕业论文，第38、39页。

〔75〕李征：《新疆阿斯塔那三座唐墓出土珍贵绢画及文书等文物》，《文物》1975年第10期。

〔76〕见CCTV-10《探索·发现》中《甘肃祁连大墓发掘记》（一），围屏似有两架，还出土有交机等，均甚罕见。

〔77〕北京市海淀区文物管理所：《北京市海淀区八里庄唐墓》，《文物》1995年第11期。

〔78〕（唐）白居易：《白氏长庆集》卷三十九"铭赞箴谣偈·素屏谣"，清文渊阁四库全书本。

〔79〕（唐）杜牧撰，（清）冯集梧注：《樊川诗集注》卷三"屏风绝句"，清嘉庆德裕堂刻本。

〔80〕（唐）张彦远著，俞剑华注：《历代名画记》卷第九"唐代上"第182页，上海人民美术出版社，1964年。

〔81〕（唐）白居易：《白氏长庆集》卷十七"律诗·题诗屏风绝句（并序）"，清文渊阁四库全书本。

〔82〕（唐）张彦远著，俞剑华注：《历代名画记》卷二，第43、44页，上海人民美术出版社，1964年。

〔83〕施爱民：《肃南西水大长岭唐墓清理简报》，《陇右文博》2004年第1期。

〔84〕传王维《高士弈棋图》（宋摹本）中亦出现坐榻后施大屏的样式，但此画榻后露出的鹤膝桌形象，目前最早见于五代王处直墓壁画，屏和榻的样式都已较《勘书图》和《重屏会棋图》成熟，应比这两幅画更晚。

〔85〕（宋）陶毂撰，孔一校：《清异录》卷下"居室·嫏宫"，第82页，辑入《宋元笔记小说大观》（一），上海古籍出版社，2001年。

〔86〕（元）伊世珍、席夫辑，（明）黄正位、黄叔校：《琅嬛记》卷中（引《丹青记》），明嘉靖刻本。

〔87〕《五山十刹图》是入宋日僧对南宋时期五大刹和十次大刹所作实录，有包括建筑、家具、法器等图样和测绘数据，详见张十庆《五山十刹图与江南禅寺》，东南大学出版社，2000年。

〔88〕唐永泰公主墓出土三彩笔架三件，座墩已经是抱鼓样式。见陕西文物管理委员会《唐永泰公主墓发掘简报》，《文物》1964年第1期。

〔89〕河南省文化局文物工作队：《河南方城盐店庄村宋墓》，《文物参考资料》1958年第11期。

〔90〕大同市博物馆：《大同金代阎德源墓发掘简报》，《文物》1978年第4期。

〔91〕目前尚未发现明代以前使用大理石屏的文献或实例，所以此屏是否确为大理石需甄别。

〔92〕河南省文物研究所、巩县文物保管所：《宋太宗元德李后陵发掘报告》，《华夏考古》1988年第3期。

〔93〕徐光冀：《中国出土壁画全集·05·河南》第129页，科学出版社，2012年。

〔94〕徐光冀：《中国出土壁画全集·10·北京 江苏 浙江 福建 江西 湖北 广东 重庆 四川 云南 西藏》第7页，科学出版社，2012年。

〔95〕徐光冀：《中国出土壁画全集·3·内蒙古》第216页，科学出版社，2012年。

〔96〕宿白：《白沙宋墓》图版5及第51页注释58，文物出版社，2002年。

〔97〕张十庆：《五山十刹图与江南禅寺》"三、家具法器"，东南大学出版社，2000年。

〔98〕四川省文物考古研究所、成都市文物考古研究所、泸州市博物馆、泸县文物管理所：《泸县宋墓》第80页，文物出版社，2004年。

〔99〕 转引自李溪《内外之间——屏风意义的唐宋转型》第四章《"文人屏"之树立》，第180页注③，北京大学出版社，2014年。

〔100〕（宋）李诫：《营造法式》上卷第六"小木作制度一·照壁屏风骨"，第127页，中华书局，2006年。"屏风骨"即屏风的木质框架。

〔101〕李溪：《内外之间——屏风意义的唐宋转型》第三章《玉堂屏风的话语之争》，第108～114页，北京大学出版社，2014年。

〔102〕（宋）陈淳：《北溪大全集》卷四"枕屏铭"，清文渊阁四库全书本。

〔103〕（宋）陈思编：《两宋名贤小集》卷七十七《潘邠老小集》，清文渊阁四库全书本。

〔104〕（宋）李幼武：《宋名臣言行录》外集卷五《邵雍》，清文渊阁四库全书本。

〔105〕（宋）陈造：《江湖长翁集》卷十八《题龚养正孩儿枕屏二首》，明万历刻本。

〔106〕（唐）白居易：《白氏长庆集》卷二十七《律诗·何处春先到》，清文渊阁四库全书本。

〔107〕（宋）苏辙：《栾城集》卷十三《画枕屏》，清文渊阁四库全书本。

〔108〕（宋）朱熹：《家礼》第五《祭礼·陈器》，清文渊阁四库全书本。

〔109〕从风格及所描绘的器具来看，此画或作于元或更晚时期。

〔110〕吕章申：《大美木艺——中国明清家具珍品》第124页"黄花梨嵌云石五屏式直腿罗汉床"，北京时代华文书局，2014年。

〔111〕（宋）赵希鹄：《洞天清录》"砚屏辨"，清文渊阁四库全书本。

〔112〕（宋）赵希鹄：《洞天清录》"砚屏辨"，清文渊阁四库全书本。

〔113〕宋一尺合今31.4厘米，据丘光明《中国古代计量史》"唐宋时期的度量衡"之"两宋度量衡量值考"，安徽科学技术出版社，2012年。

〔114〕（宋）李诫：《营造法式》上卷第六"小木作制度一·照壁屏风骨"，第127页，中华书局，2006年。

〔115〕田洪：《王季迁藏画集》"宋元编"，天津人民美术出版社，2008年。

〔116〕（宋）欧阳修：《欧阳文忠公集》外集卷十五《月石砚屏歌序》，四部丛刊景元本。

〔117〕（宋）欧阳修：《欧阳文忠公集》居士集卷第四《紫石屏歌（一本作〈月石砚屏歌寄苏子美〉）》，四部丛刊景元本。

〔118〕（宋）吕祖谦：《宋文鉴》"皇朝文鉴卷第十三·永叔月石砚屏歌"，四部丛刊景宋刊本。

〔119〕（宋）杜绾：《云林石谱》卷中"虢石"，清文渊阁四库全书本。

〔120〕苏轼在《书月石砚屏》中还提到月石屏的鉴别方法，"月石屏，扣之，月微凸，乃伪也。真者必平，然多不圆。圆而平，桂满而不出，此至难得，可宝"。此说概只适合小型石板，较大的石板加工甚难平整，所见古屏石板面多凸凹，这反而是今人鉴别真品标准。再者，石板纹路多是匠师利用石板不同层的色差，磨制高低以显现不同颜色纹路，追求虽由人作宛自天开的效果，其实欧阳修亦注意到这一问题，"不经老匠先指决，有手谁敢施镌镵"。可见苏轼所言也不尽然。

〔121〕（宋）陶穀撰，孔一校：《清异录》卷下"陈设·玉罗汉屏"，第93页，辑入《宋元笔记小说大观》（一），上海古籍出版社，2001年。

〔122〕王清林、周宇：《石景山八角村金赵励墓墓志与壁画》，辑入北京市文物研究所《北京文物与考古》第五辑，北京燕山出版社，2002年。

〔123〕徐光冀：《中国出土壁画全集·02·山西》第135页，科学出版社，2012年。

〔124〕本文所述大理石为传统文化概念中的大理石，与地质学中大理石的概念并不等同。

〔125〕（明）文震亨著，陈植校注：《长物志校注》卷六"几榻·一九、屏"，江苏科学技术出版社，1984年。

〔126〕（明）兰陵笑笑生著，陶慕宁校注：《金瓶梅》第45回，第529页，人民文学出版社，2008年。

〔127〕蒋晖：《明代大理石屏考》第一章"明早期·一 格古寻石"，山东画报出版社，2018年。

〔128〕（明）郎瑛：《七修类稿》卷四十四"事物类·料丝"，明刻本。

〔129〕（清）姜绍书：《韵石斋笔谈》卷下"丝灯记略"，辑入《美术丛书》三集第一辑。

〔130〕（明）陆深：《俨山集》卷二，清文渊阁四库全书本。

〔131〕收入故宫博物院编《故宫博物院藏明清家具全集·20·屏 联》件133，第506～507页，故宫出版社，2015年。

〔132〕《观台磁州窑址》曾收录一件南宋景定三年款的白地黑花瓷屏，但实物今不知何处。转引自扬之水《终朝采蓝——古名物寻微》"砚山与砚屏"注18，第195页，三联书店，2012年。

〔133〕王世襄：《明式家具研究》附录"美国加州古典家具博物馆"，第402页，三联书店，2011年。

〔134〕收入故宫博物院编《故宫博物院藏明清家具全集·20·屏 联》件104，第360～365页，故宫出版社，2015年。

〔135〕如故宫博物院编《故宫博物院藏明清家具全集·19·屏 联》件51"紫檀嵌珐琅玉山水人物图鱼缸插屏"，第144～145页，故宫出版社，2015年。

〔136〕（元）施耐庵、（明）罗贯中：《水浒传》第七十二回《柴进簪花入禁院，李逵元夜闹东京》，第1061页，容与堂本，上海古籍出版社，1988年。

〔137〕（清）李周望辑：《国学礼乐图》卷之十五"礼器图·祝版"，清康熙五十八年国子监刊本。

Seat Right
A Brief Talk on Ancient Chinese Seating Screen

Zhang Zhihui

Screens are a unique category of furniture in China, with a long history. Screens appeared as early as the Zhou Dynasty and were the product of civilization reaching a certain stage of development rather than a necessity of daily life. The decorative, partitioning, and shielding functions gradually developed over time.

Based on form, screens can be divided into two main categories: table screens and folding screens. Table screens are those on a base and this were the earliest type. Folding screens are multi-panelled, free standing, connected by hinges and other mechanisms. This article primarily focuses on table screens, presenting the history of their development in chronological order.

During the Zhou Dynasty, there were two important types: Fuyi and Huangdi. The emperor sat facing south, and the Fuyi was placed to the north of the Ming Hall, behind the emperor, symbolizing the emperor's power and status. The Huangdi was a panel screen decorated with bird feathers and used during ritual ceremonies to the Heavenly Emperor, symbolizing the divine entity.

In the Warring States period, an important example was discovered in the tomb of the King of Zhongshan. They included bronze screens with gold and silver tiger devouring deer motifs, bronze screens with gold and silver rhinoceros' motifs, and bronze screens with gold and silver ox motifs, all exquisitely crafted.

During the Han Dynasty, screens became not only ceremonial and shielding objects but also symbols of splendid power and status. When displayed indoors, their form and scale varied depending on the user's social status. Figures of sages, virtuous women, and other characters appeared in the patterns, serving instructive and educational functions.

In the Sui and Tang periods, large screens began to be widely used. The six-panel folding screens, used as carriers of calligraphy and painting, also became prevalent. Abundant images of six-panel folding screens can be seen in literature, tomb murals, and examples from this period.

During the Song Dynasty, screens had developed a systematic structure and matured in terms of structure and form. Additionally, small and medium-sized screens such as pillow screens and inkstone screens emerged.

Pillow screens were extensively used during the Song Dynasty and had some connection to the use of beds and couches outdoors or in pavilions. Placed on the side of a pillow, they could provide wind protection.

Inkstone screens were another important type of screen that emerged during the Song Dynasty and were related to the trend of literati appreciating inkstones and rare stones. Notably, Ouyang Xiu's Moonstone Inkstone Screen gained fame, and Ouyang Xiu, Su Shunqin, Mei Yaochen, Su Shi, and Su Zhe wrote renowned poems about it.

In the Ming Dynasty, especially in the mid-Ming period and beyond, with the development of the small commodity economy, hardwood furniture, led by *huanghuali*, became popular. Delicate and refined hardwood furniture competed with free-spirited softwood furniture, reaching the pinnacle of high-class Chinese furniture. It encompassed a wide range of categories, matured in structure, varied in decoration, and featured intricate craftsmanship.

The screen examples seen today mainly consist of small and medium-sized placed on desks during the Ming and Qing dynasties. Ming Dynasty screen frames were often lacquered, while in the Qing Dynasty, they were made of hardwood such as *huanghuali* and *zitan*, with screens made of stones from Dali and Qiyang.

图版
Plates

长河遗珠

Treasures from the Long River of History

001

黑漆薄螺钿
孔子观欹器图座屏

明中晚期

高 67.5 厘米　宽 72.1 厘米　厚 29.6 厘米

此例为中大型座屏，横屏式，宽72.1厘米，非大案不足以承，也有可能是枕屏之属（关于枕屏，详见前文《中国古代座屏概谈》）。楠木胎，素黑漆，漆、灰薄而致密，小细碎断纹如鱼鳞，间有个别发丝般长断纹。

正面嵌薄螺钿《孔子观欹器图》，典出《荀子·宥坐》：

孔子观于鲁桓公之庙，有欹器焉。孔子问于守庙者曰："此为何器？"守庙者曰："此盖为宥坐之器。"孔子曰："吾闻宥坐之器者，虚则欹，中则正，满则覆。"孔子顾谓弟子曰："注水焉。"弟子挹水而注之。中而正，满而覆，虚而欹。孔子喟然而叹曰："吁！恶有满而不覆者哉！"

欹器的造型为尖底瓶状，或言源自古时汲水陶器。欹器中部有耳，系于架上，注水满则头重脚轻，上部倾倒而水溢出，警示观者要时时谦逊，戒自满。传古时君王常置欹器于座右，用以自警，故又名"宥坐"，孔子所见即鲁桓公之宥坐。故宫博物院藏有清铜鎏金带座欹器。

画面左侧有垂花门、照壁，意鲁桓公庙。欹器三件，居右者略斜，"虚则欹"；居中者竖直，注水恰到好处，"中则正"；居左者还在注水，即将倾倒，"满则覆"。欹器固定于架，下附须弥座。座束腰饰卷草纹、海马纹。孔子位于画面中心，形体大而突出，符合所谓身长九尺二寸的"长人"特征，开脸亦很形象，环眼、隆鼻、骈齿、大耳，与历代传画之孔子像相若。孔子对面老者当为守庙人，其身后四人及注水两人为其弟子，尚有汲水的童仆一。

画面构图严谨，嵌饰工细，如同界画，一丝不苟，制作者极着力于画家笔下物象的准确表达，从笔墨到螺钿镶嵌工艺之间的转换几无差别。一般所见薄螺钿镶嵌，多有浓郁的装饰性，稿本多取布局较满者，以形成纷繁复杂的效果。此屏不然，虽然主体人物衣物以各式锦地装饰得繁华旖旋，但布局疏朗，简繁对比明显，物象醒目。留白的大片黑漆因年久形成的皮壳如水、如天、如云，令人遐想不断。薄螺钿镶嵌方法也与同类有别，尤其是栏杆、衣物等处，采用了模切般小单元螺钿片攒成锦地的装饰方式，布局工整细腻，图案变化多端，加以其上划出的细小纹路，光线浮动，甚觉华丽精致。日本、朝鲜螺钿工艺中有一种以长方条垫片填充的方法，原理接近，但效果呆滞粗鄙，远逊此屏。

屏背以厚螺钿嵌成行书："积金以遗子孙，子孙未必能守。积书以遗子孙，子孙未必能读。不如积阴德于冥冥之中，以为子孙长久之计。此先贤之格言，乃后人之龟鉴。"此语见司马光《家训》，为中国古代经典格言。

《孔子观欹器图》是较为经典的历史故事，绘画、版画皆有表现，如孔子博物馆藏有明代所绘者，布局、风格与本屏皆相近，以之装饰于漆器或家具上，却是罕见。孔子是古之圣人，欹器是古代贤君座右之器，故此题材较为严肃，鉴戒意味甚浓，饰于屏者，应是特殊用具，或为帝王宫殿所陈，或为孔庙、学府等地所设者，非一般日常家用之器。屏背嵌厚螺钿的书体和形式，是明末清初时较为流行者，虽亦属难得，但境界和意趣远逊正面，缘何如此，尚待探讨。屏风初为礼仪之具，至迟汉时已具鉴戒之器的属性。西汉羊胜《屏风赋》有"画以古列，颙颙昂昂。藩后宜之，寿考无疆"句。刘向与子刘歆以"《列女传》种类相从为七篇，以著祸福荣辱之效，是非得失之分，画之于屏风四堵"，皆是此类。新疆阿斯塔那216号墓出土壁画所绘六曲围屏，中间四扇为"石人""金人""土人"等人像，右边扇为扑满、丝束和青草，左边扇即为欹器，均有明显的鉴戒意味。这也是所知较早在屏具上饰欹器图案的实例。

屏框上方有委角，下宽上窄，侧脚明显。座下设壶门牙板，曲线柔和悠扬。屏墩抱鼓式，下开壶门亮脚，两头伸出，弯钩状。壶瓶式站牙。

此屏延续宋代座屏样式，薄螺钿镶嵌图案为明中晚期风格。

63

積金以遺子孫,子
孫未必能守積書以
遺子孫,子孫未必能
讀不如積陰德於
冥冥之中以為子
孫長久之計此先
賢之格言可後人
之龜鑑

Mother-of-pearl inlaid black lacquer table screen

Middle to Late Ming Dynasty
H 67.5CM W 72.1CM D 29.6CM

Of rectangular form, inlaid with mother-of-pearl on black lacquer. Obverse side depicts a story, Confucius (551BCE–479BCE) and his students in the garden of the Ancestral Temple studying the water equipment during his visited at LU Kingdom.

Confucius stands in the middle of the scene and is of a slightly larger proportion whilst talking to the Temple Guardian. Two assistants are pouring water into a bucket preparing equipment for study. A small servant draws water from the well in the lower left corner. The reverse side is inscribed in semi-cursive script with an ancestral aphorism. The shape of frame is slightly narrower width in the upper portion compared to the lower. The scrolling curved apron attached.

The subject of the decoration on this screen is related to Confucius and ritual infers the use of this piece is for formal placement or occasions, not for daily causal actives. The inlaid mother-of-pearl techniques used on the script was a popular method during the late Ming and early Qing period. This technique was also used for decoration on lacquer objects by Japan and Korean during the same historical period.

002

黑漆薄螺钿佛像图座屏

明中期

高 67.5 厘米　宽 83 厘米　厚 28.5 厘米

　　座屏尺寸较大，制式高古。

　　屏心薄螺钿嵌成佛像图，波浪滔天，浮有三山。中山较高，其上释迦佛跏趺坐，着袈裟，袒右臂，施说法印，有头光和背光。余两山上为文殊与普贤二菩萨，一持莲花，一持经卷，跏趺坐于狮、象上。佛和菩萨的服饰皆以菱形螺钿片排列镶嵌，并间饰圆螺钿片或花状螺钿片，佛像选黄绿色螺钿片，菩萨像选紫色螺钿片，珠光宝气。前方海面之上，为十八罗汉渡海图，每个罗汉皆以云纹界出如开光状空间，各显神通，自右向左而行，略成三列，乘莲蓬、锡杖、龟、象、蒲团、鳌鱼、海马、杖、树叶、犀牛等宝物，还有数个罗汉尚在岸边将渡。对面为海中龙宫，露出歇山式殿宇一角，画面左下方龙王手持笏板，躬身迎接，旁有从官、卫士。佛、菩萨皆稍低首观看罗汉渡海。

　　罗汉渡海图属佛道画，在宋、元、明时期甚为流行，多为长卷，表现波涛之汹涌，罗汉之镇定悠闲和超群法力。此屏图案更加复杂，增添了上方的一佛二菩萨，且形象更加高大醒目，与罗汉一同成为主题。

　　屏心为一块漆板，背附穿带，可活拆，在背后上方横枨上设两个活销固定。座屏中屏心为箅子者，上裱书画，常作成活拆式，此例为将漆板做成活拆者，甚是少见。

　　复框式，上窄下宽，挓度明显，更显稳重安定。外框转折处外侧做成委角，内侧做一个小角如钩，颇为别致。复框间绦环板漆地上嵌薄螺钿为球路纹锦地，中间点缀栀子花，锦地上有圆形开光，内饰莲花，与屏心佛像主题呼应。绦环板转角处饰有如意云头，这种装饰在宋元绘画中的屏风上常可见到。屏框素黑漆，发色纯正，上有蛇腹断。屏背光素，薄擦黑漆而成。

　　屏下牙板高古，为连弧锯齿状，弧尖做小委角，与魏晋佛像中常见的壸门样式相类，应亦是其设计思想的源头。另需注意的是常见明清座屏的下方横枨多与屏墩上沿平齐，下装牙板，亦有连接屏墩的作用。此处横枨高于屏墩，牙板装在屏墩之上，下方留空较多，更觉高挑，概是沿袭早期做法而来。

　　站牙为如意云头式。屏墩抱鼓墩式，抱鼓上亦饰莲花，其下雕为草芽，屏墩下两足撇出，末端翻卷为花叶状，中有壸门式亮脚。站牙、抱鼓墩皆甚窄小紧凑，与宋时图像所见座屏比例相近。

　　屏上以菱形或条状螺钿片排列嵌成物象的做法，与常见的薄螺钿镶嵌不同，有观点认为这种属于日本或琉球风格。从传世的实物来看，确有这样一批地域难以确定、宋元风格甚浓的薄螺钿器物，既与典型中国风格不同，亦与日本传统漆艺迥异，似乎介于两者之间，具体若何，尚需更深入的实物总结和实地考察。以此屏而言，其造型、漆质乃至榫卯结构，更具中国特征。

　　屏心装饰，以山水为大宗，其次有花鸟、瑞兽、人物故事等，以佛道图像为装饰者，甚为罕见，一般为寺院或者信徒供养所用。从屏的源流来看，起初有周天子所负之黼依和祭祀上帝的皇邸，佛道主题类屏属于后者遗意。

　　此屏造型典型宋式，漆片采样经碳十四检测为1478年±21年，可资断代参考。从家具史来看，明代家具风格的确立在明中期，此前的家具依然沿袭宋元风格，即如此屏。

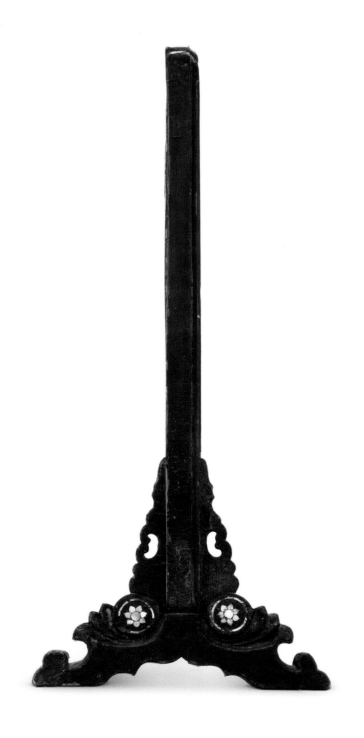

MOTHER-OF-PEARL INLAID BLACK LACQUER TABLE SCREEN

MIDDLE MING DYNASTY
H 67.5CM W 83CM D 28.5CM

The subject depicted is *Eighteen Luohans crossing the sea*.

Large size presented in the ancient artic style. Single lacquer work used as central panel, there are two removable lockers attaches to the back of central panel which is a rare example of this technique used during the period.

Four rectangular panels surround a central panel with the top panel narrower than the bottom panel employing a mother-of-pearl inlaid on black lacquer technique. The frame using pure black lacquer with surface slightly cracked. The back of the frame has a thin layer of the black lacquer.

Flower patterns decorate the top of a dense geometric background. The lotus flowers depicted on the frames are related to a Buddhist topic on the central panel. The carved panel has a cloud shaped header. This decoration method is commonly seen on paintings from the Song and Yuan dynasties.

The scalloped shaped apron under the panel is attached to the top of the stand, a rail (side stretchers) between apron and stand. The screen stands presented as a drum shape and decorative with lotus flower insert.

The diamond and stripe shaped mother-of-pearl inlaid techniques which are used on the screen may relate to the Japanese or Ryukyu lacquer technique.

73

003

黑漆薄螺钿
雅集图座屏

明末清初

高 52.5 厘米 宽 63.5 厘米 厚 22.2 厘米

中型座屏，造型、纹饰皆延续宋元样式，黑漆薄螺钿装饰。

屏心为雅集图，庭院中一处楼阁，楼下四人正在品评一幅山水画卷，隔扇的另一边有三人私语。楼上三人对坐，似在观书。楼旁水榭中，一人弹琴，二人聆听，一人斜倚栏杆，探身观看水面莲花。庭院中三人相与而来，童仆相随，一只白鹤，昂首舒翅，似乎在欢迎并指引到来的三人。楼阁旁一株古松，水榭旁植垂柳。

雅集图是自宋以来甚为流行的图画内容，著名者有《十八学士图》《西园雅集图》等，多表现文人雅士在庭院中从事琴、棋、书、画等活动的场景，在漆器、瓷器、玉器、家具等工艺美术中亦多有采用，以彰显使用者的雅趣。

整个画面布局较满，多以细碎的条状螺钿片镶嵌图案，体现稿本的白描效果。人物的脸部、幞头及仙鹤等则以大块螺钿片镶嵌，再以阴线勾勒图案而成。衣物等以三角形、花形等样式的螺钿片排列填充而成。

复框式，托度明显，边框看面起剑脊棱，线脚较繁。复框间装绦环板，上、左、右连为一体，转折处有如意头状装饰，下方并列两个绦环板。绦环板球路纹锦地上设菱花形开光，上方为凤穿牡丹纹，两侧为童子持莲纹，下方为对称的缠枝灵芝纹与奔兽纹。图案风格高古，皆宋元时期样式。嵌饰则螺钿条与螺钿片间施。

屏下施断牙式花牙，有以红漆髹饰的S形花叶。屏墩做法较为别致，在屏框外另设立柱，形如插屏结构样式，其实屏框与立柱连为一体，不可拆卸。石榴头式站牙，曲线饱满，髹红漆装饰。屏墩抱鼓式，抱鼓亦髹红漆，较小巧，下为花叶形造型，两足撇出。

此屏之造型、纹饰与工艺皆有早期特征，然漆片经碳十四检测，年代参考为1633年±22年，与初步之判断相差甚多，这提醒我们在断代时需慎之又慎。同样风格的薄螺钿制品见有数例，亦有观点认为是琉球制品，均暂缺乏深入的研究，如果以此屏为起点，对这些漆器进行详细的归纳总结，并辅以科学检测，相信会对这一具有特殊地域、时代风格的薄螺钿漆器有更多新认识。

75

Mother-of-pearl inlaid black lacquer table screen

Late Ming to Early Qing Dynasty
H 52.5CM W 63.5CM D 22.2CM

The subject depicted is *Elegant gathering*.

The central panel depicts scholars or officials studying the antiquity or chatting together in the garden. Scholars gatherings were a popular theme for decoration since the Song Dynasty.

The mother-of-pearl is cut very thinly, fine and narrow elongated strips inlaid into the black lacquer. The colour contrast between black lacquer and the mother-of-pearl heightens the colourful outline effect. Larger pieces of mother-of-pearl inlay are used for people's faces, headdresses and crane's body. This inlay technique suggests similar characters to lacquer works from the Ryukyu Kingdom (1429-1879). Future discussion of this opinion requires further academic research.

Carbon-14 tested on fragments of lacquer, dated 1633 ± 22 CE which places it in the Chongzhen period.

004

黑漆薄螺钿
人物故事图座屏

明万历

高 48.3 厘米　宽 44.8 厘米　厚 26 厘米

屏心黑漆嵌薄螺钿而成。外围为锦地，上有开光，如同画幅的裱边。锦地为六方形攒成，中为栀子花纹。菱花形开光，内为薄螺钿花鸟图案，花为芙蓉、石榴、林檎之类，鸟皆成双。开光内图案嵌饰极为工整，与一般装饰不同，为明中晚期小写意花鸟画风格，应是根据名手画稿而成，其中上方的芙蓉孔雀图案，在当时漆器、瓷器等装饰中甚为流行。

画面为湖畔一景，以右下角至左上角的坡岸将画面分为两部分。左下侧为近景，一处高台上有阔大厅堂，槛窗饰以锦地，上设太师窗，螺钿片嵌成的柱子上布满图案，阴刻童子抱莲纹，这是宋元以来建筑上常见的装饰手法，诸如苏州罗汉院正殿遗存，柱身满刻童子缠枝花纹，精彩绝伦。厅堂的门帘卷起，上密饰缠枝花纹。正中设大供桌一，杂陈表册。桌面嵌为锦地，周匝饰香草纹，三弯腿，设壶门牙板，下承托泥。后设地屏，大宽边，屏心为锦地，亦为六方形，中有栀子花。台基上围以锦地栏杆，一男子身着华服，满饰云鹤纹，斜倚栏杆，正在观看院内。一旁童子横持羽扇，屏侧有仕女走出。台阶下四个童子，前后相随，手持盒、钵、琴等物，有仙鹤衔卷轴起舞。厅堂旁一株梧桐夐绕而生，画面近处围以栏楯，锦地为饰，栏楯周匝有菊花、箸竹、芭蕉等植物。右上侧为远景，山岩上杨柳依依，两只飞鸟划过半空，山岩下露出半叶扁舟，一童子赤腿坐于船头，右脚点在水面上，荡起一片涟漪，双手横笛鸣奏，笛声若闻。

此屏画面内容奇特，主题成谜，后见康熙版《历代神仙通鉴》书前版画，所绘神仙梅福与陈生，与此处相像，尤以陈生形象，和屏中奏笛童子几无二致，则知此屏与以前盛行于江西、安徽、江苏、浙江、福建等地的梅仙信仰有关。梅福为西汉末九江寿春（今安徽寿县）人，精通儒家经典，汉成帝时为南昌尉，预见王莽之祸，辞官避祸，入山修仙。陈生善笛，随梅福修仙，后窃食仙丹，逃入浮石洞，每至风月之夜，渔舟泊者，闻笛声隐隐自水中起。

屏框帐子正面为打洼线脚，背面中起剑脊棱，两边起阳线。复框式，装一周绦环板，上为长海棠式开光，开光内两头镂雕为如意云头状，上、左、右三面开光中间镂为网格状，空灵疏朗，下方开光中间镂为剑环套双方胜纹，富于变化。屏下设壶门披水牙板，垂肚较大，两头卷云纹相抵。变体壶瓶式站牙，上饰卷草纹。抱鼓式屏墩，扁球状抱鼓，甚为饱满，其下两层，上层托住扁球，出头处雕为卷珠，下层若几座，三弯腿外翻马蹄，中间亮脚为壶门式。屏框、座的样式可以上溯至宋，但帐子的线脚、牙板的壶门曲线和相抵的卷云纹样式，都已有明代特征。

81

屏心画面构图考究，嵌饰法度谨严，物象准确，生气勃勃，非高手难以为此。多采用较大钿片上勾勒复杂线条表现物象的方式，既保持整个画面的整洁，又达到了装饰的需求。尤为精细的是画面中的石头纹路里，刻画各种动物图案，如殿宇旁梧桐树下的四块石，大者内刻卧鹿、飞鸟，余处为卷草纹；次大者内刻灵芝卧兔，两小石内亦含物象，较抽象简略。近景处栏杆旁三块石，一块螺钿片已失，中间一块上刻蚊虫、三足蟾。栏杆内有小石，内刻鸟雀。坡岸处的石上刻有飞虫，水中石上刻水鸟。螺钿工艺中这种表现石头的手法，甚为少见，或是要表达石中皆含灵异，这实际是一处神仙府邸所在，正与前述梅仙信仰相合。

屏背嵌厚螺钿行书词句："龙楼凤阁九重城，新筑沙堤宰相前。是一书生行我贵，我荣君莫羡十年。"此诗句出自马致远杂剧《半夜雷轰荐福碑》，为剧中范仲淹所念，明清时作为格言甚为流行，但后两句原为"我贵我荣君莫羡，十年前是一书生"。此处将字顺序嵌错，甚至将原来"宰相行"的"行"字与后面的"前"字嵌混，想是工匠不通文墨，嵌时将文字顺序弄错。螺钿字下有墨笔字，虽已残损不全，但从遗留的字迹可以较容易辨认内容是明初人陈献章《忍字赞》：

> 七情之发，惟怒为遽。众逆之加，惟忍为是。绝情实难，处逆非易。当怒火炎，以忍水制。忍之又忍，愈忍愈励。过一百忍，为张公艺。不乱大谋，其乃有济。如其不忍，倾败立至。

此处为行书小字，远比厚螺钿嵌成文字娴熟练达，可推知原为此箴言，后又改嵌厚螺钿词句。座屏后背饰箴言文字，是屏具作为鉴戒之具的原始用途遗存，本书收录黑漆螺钿孔子·观欹器图座屏，背后嵌饰司马光《家训》，亦是此类。

此屏造型上承宋元，嵌饰图案精彩绝伦，繁而不乱，保存甚完整，发现自东邻日本，但嵌饰手法、图案等，都为典型的中国明晚期样式，应为当时传入日本者。漆片经碳十四检测，为1599年±21年，即明万历时期，与判断年代相符。相较于本书所录的同时代红漆框镶绿石座屏，造型稍复杂些，然对比两者，依然可以发现相通的时代气息和若干接近的细节处理。

Mother-of-pearl inlaid black lacquer table screen

Wanli Period, Ming Dynasty
H 48.3cm W 44.8cm D 26cm

The central depiction is of a scholar with his two attendants seated inside a pavilion. Four young servants bearing instruments and boxes seem to walk into the pavilion. Immortals may inhabit this place as the flying and water birds are carved in the rock which would indicate the magic power of these rocks. The whole painting manages the space very well, exemplified by the clever logical order of the positioning of the landscape and buildings. These details suggest this painting may have been selected from a famous artist's drawing during the period.

The inscription on the back cited from Yuan opera *Banye Leihong Jianfubei* (*Thunders Cutting Jianfu Stele in the Midnight*) written by Ma Zhiyuan (1250-1321). Within this paragraph, two characters are different from the origin text, which may indicate the craftsman may not have read or understood the text.

Carbon-14 tested on fragments of lacquer, dated 1599 ± 21 CE, which is Wanli period.

005

黑漆薄螺钿
三仙贺寿图插屏

清早中期
高 55.3 厘米　宽 39.5 厘米　厚 20.5 厘米

　　竖屏式，屏心只上部镶框，竖框搭在屏座的立柱上。

　　屏心边框打洼，撒五彩螺钿沙为地，别出心裁地嵌大小银片为梅花，花心以螺钿点缀。屏心薄螺钿嵌成三仙贺寿图，红日当头，滔天波浪拥簇三座仙山，山上仙草生发。三山中间高而旁边低，有拱卫之意。水纹以细如发丝的细螺钿条嵌成，浪花以银片嵌成，仙山之后的远山为撒螺钿沙而成。右下坡岸，不同位置地面以红、棕、绿三色螺钿沙撒成，其上古松夭矫，湖石玲珑，仙鹿抬首探身，瑞鹤空中回旋。松下三仙望山而贺，持杖隆额者为寿星；捧桃文士或为东方朔；近处背对观者的仙人手举卷轴，大袖飘舞，或为控鹤仙人，据《列仙全传》载其常控鹤，曾至武夷山校定仙籍。近景处二童子捧宝物走向三仙。屏背撒螺钿地上嵌银片五岳真形图。

　　屏座设素黑漆绦环板，上设剑环式开光。站牙为壶瓶式，抱鼓式墩。屏座横枨上薄螺钿折枝花卉纹，为石榴、佛手、桃"三多"纹，含祝寿寓意。壶门披水牙板饰折枝石榴花与牡丹花。

　　黑漆质地细腻，几不见断纹，发色纯正，螺钿精巧细致，五彩斑斓，均为清早中期漆艺特征，与故宫博物院所藏数件康熙款漆家具接近。其二山拱卫一山、三仙遥祝、一日当天的图案，似乎也有朝拜天子之意，而三山瘦挺的海水江崖纹、五岳真形图都是雍正时期宫廷器具常用的图案，不排除此为雍正或乾隆初期宫廷用具。

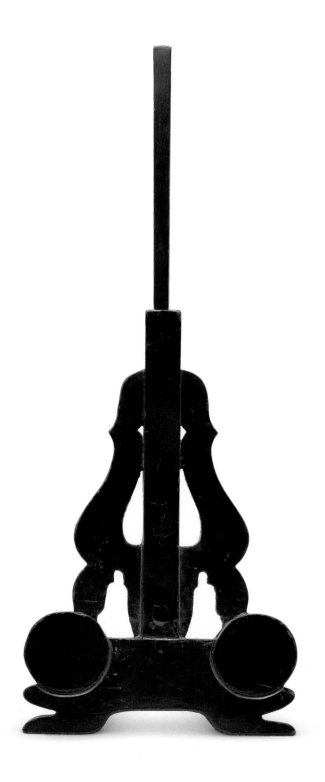

MOTHER-OF-PEARL INLAID BLACK LACQUER TABLE SCREEN

EARLY TO MIDDLE QING DYNASTY
H 55.3CM W 39.5M D 20.5CM

The subject depicted is *Three-star gods worship scared mountains.*

The special technique on this piece are thin fine cut and minute grounded mother-of-pearl inlaid on top of black lacquer. This thin cut mother-of-pearl brings rich colour which, under the lights, can enhance the Daoist magic power of the subject, such as the waves of the sea between the three gods and scared mountains.

The True Forms of Five Sacred Mountains on the back of central panel is one of ancient Daoist scripts protecting people from getting lost when passing the mountains and also preventing attacks from evil spirits. Most importantly, it could bring good fortune. True Forms of Five Sacred Mountains are among the most favoured decoration patterns during the Yongzheng period (1723-1735).

006

黑漆薄螺钿缠枝花纹框镶红漆诗文座屏

明晚期

高 51.5 厘米　宽 58.5 厘米　厚 19 厘米

　　横幅式中型座屏，宋元风骨尚存。

　　屏心红漆地，正面描金行书诗文："绛苞迎腊小寒前，看到春初破未全。难进自坚君子节，争先应避艳阳天。偶沾宫额原无意，独伴山窗似有缘。茅屋玉堂清绝处，癯然风骨是神仙。"落款"莆阳蔡襄"，绘朱文印"太函""玄扈司马"。背面墨绘山水，辅以金色，绘竹石秋树，颇为萧疏，意近倪瓒。查此诗并不见于蔡襄诗集，只见于清康熙华希闵等刻本《朴村诗集》的注解提到"蔡君谟《十梅诗》有'绛苞迎腊小寒前，看到春初犹未全，难□自坚君子节'云云，先生年近六旬，方魁天下"。清抄本陶越《过庭纪余》记明成化间金溪徐太守蒎禾有《梅花十咏》："八曰：绛寒（此字圈过）迎腊小寒前，看到春和破未全。难进似坚君子节，争先应避艳阳天。偶沾宫额元无意，独伴山翁似有缘。茅屋玉堂清绝处，癯然风骨似飞仙。"与此屏文字基本相同。明代歙县名士汪道昆（1525～1593年），字伯玉，号南溟，又号太函，文武兼备，与王世贞为当时诗坛领袖，著作颇丰，且精通音律、戏曲创作，声名远扬。著作有《太函集》《玄扈楼集》等，"玄扈司马"印则见钤于汪道昆在台北故宫博物院藏周天球《墨兰图》卷的题跋处。所见汪道昆法书，较为瘦劲，然此书风较为丰腴，应为赵孟頫一派，题者当另有他人，或是明人根据有汪道昆印鉴的蔡襄款法书，移摹而成。

　　复框式，上方带委角，托度明显。框皆细挺，正面起剑脊棱，背面平直，髹黑漆，漆质细腻，发小蛇腹断。绦环板薄螺钿嵌成锦地，为六边形套栀子花式，其上有菱形开光，内薄螺钿嵌成缠枝花纹，繁密细致，枝叶勾卷，花头有菊、莲、石榴、牡丹、栀子、山茶等类，延续元代螺钿漆器装饰手法。背面绦环板则描金饰折枝梅花、荷花飞鸟。

　　屏框高脚，下方有素直牙板。抱鼓墩比例秀巧，抱鼓小，下方托以花叶，有高亮脚，两足若蹄，制式高古。桨腿式站牙，质朴简练。

93

Black lacquer framed inlaid with cinnabar lacquer table screen

Late Ming Dynasty
H 51.5CM W58.5CM D: 19CM

Central panel coated with cinnabar lacquer containing a poem written in semi-cursive script.

Double frame style, with chamfered shoulders on top panel. The frames are solid and straight, with sword ridges. The back is black lacquer with fine lines resembling snake belly crazing.

The plate is inlaid with thin mother-of-pearl into a brocade background, which is a hexagonal gardenia pattern, with diamond-shaped openings.

The inner thin mother-of-pearl is inlaid into a vine like pattern, which is dense and delicate, with curling tentacles and leaves. The flower heads have chrysanthemum, lotus, pomegranate, peonies, gardenias, camellias and other, continuing the Yuan Dynasty mother-of-pearl lacquer decoration technique.

The plate on the back is decorated with *miaojin* technique, depicting plum blossoms, lotus flowers and birds.

The screen frame has high feet, and there is a plain straight apron attached. The proportion of the drum shaped stand is exquisite, with flowers and leaves beneath, with simple and clear hooved footer of an ancient style with a simple paddle-leg stand.

許並迎臉小寒芳看列秦
初破朱金蕊逞冷艷君子
許爭先夜避艷陽天偶沾
宮額尔無意禍任山冠松
有緣芳屋玉堂清陰
雍然風骨見丹仙

007

黄花梨螭龙纹框
黑漆刻灰水仙图插屏

明末清初

高 57.5 厘米　宽 55.5 厘米　厚 19.5 厘米

屏心为一幅花卉小品，黑漆刻灰水仙一株，叶肥花丰，摇曳生姿，衬以竹枝，落款"陈继儒书"，刻"陈继儒印""眉公"二印。

陈继儒为明晚期极负盛名的文人隐士，与董其昌齐名。"当启、祯间，妇人竖子，无不知有眉公者，至饮食器皿，悉以眉公名。比于东坡学士矣。"（计六奇《明季北略》）"上自缙绅大夫，下至工贾倡优，经其题品，便声价重于一时。故书画器皿，多假其名以行世。"（黄宗羲、黄垕炳《海外恸哭记》）"近而酒楼、茶馆，悉悬其书画，甚至穷乡小邑，鬻粔、妆市、盐豉，自床、几、椅、杌，以至一巾一履，胥被以眉公之名，无得免焉。"（张岱《石匮书》）

此屏而言，当可以"眉公屏"名之。陈继儒能书善画，传世作品较多，水仙主题的绘画也见有数幅，故宫博物院藏有《竹仙梅花图》册页，其中一幅绘水仙梅花，与屏心水仙风格趋近，可知屏心蓝本应出自眉公绘稿。唯落款从内容看，似辑自陈继儒法书，并非原画原跋。

刻灰即款彩工艺的一种，只是刻灰雕刻后不再填色，形成拓片般素雅的效果，多受文士青睐，应用不甚广，见于江南地区的一些文玩器具之上。

屏框宽边厚料，颇稳重。屏座上设绦环板，剑环式开光内透雕双螭戏珠纹。螭为俯视正面像，首略长，扁嘴，杏核眼，独角，发两束，兽身有背鳍，三趾式四足，卷草式尾，作折身回望式。这种样式的螭纹在清早期器物应用甚多。披水牙板壸门式，中间柿蒂纹，两边衍出卷草纹。变体夔龙式站牙。抱鼓墩，下开壸门，中饰简练的如意纹，正中起竖脊以增层次。黄花梨框、座皆灰褐色皮壳，甚为雅洁。

此屏屏心刻陈继儒《水仙图》，是否当时制作不得而知，从螭龙纹样式等看，应制于康熙时期或更早。

屏背黑漆刻灰书法六句，依次为："曲江贵家游赏，则剪百花妆成狮馈遗，狮子有小连环，欲送则以蜀锦流苏牵之，唱曰：'春光且莫去，留与醉人看。'""张翔好学，多思致，尝戏造《花经》，以九品九命升降次第之。""陶弘景特爱松风，庭院皆植松，每闻其响，欣然为乐。""洛阳梨花，时人多携酒树下，曰：'为梨花洗妆。'""房寿六月召客，捣莲花制芳酒。"落款为："冬日雨窗王阊书。"刻"王阊"朱文长方印。皆为涉花树的雅事。行书，前五句较工而近楷，笔画瘦劲，第六句近草，略丰腴。第一、四、五句出自后唐冯贽《云仙杂记》，据载分别引自《曲江春宴录》《唐余录》《叩头录》，第二句出自宋陶穀《清异录》，第三句出自唐李延寿《南史》。王阊，字季仙，嘉兴人，活跃于明晚期。《长洲文氏尺牍册》（故宫博物院藏）《李流芳山水册》（《别下斋书画录》）有其题跋及钤印。

明
陈继儒
《竹仙梅花图》册之一
故宫博物院藏

Ming Dynasty
Chen Jiru
One of Album of *Bamboo, Narcissus and Plum Blossom* (Zhuxian Meihua Tu)
The Palace Museum

Huanghuali framed *kehui* black lacquer table screen

LATE MING TO EARLY QING DYNASTY
H 57.5CM W55.5CM D19.5CM

The central panel depicts Chinese daffodils which have auspicious meaning as a water immortal. The inscription shows 'Chen Jiru shu'. Chen Jiru was famous for calligraphy and flower paintings. The work on the central panel of the table screen uses the same technique as the brush work from Chen's work *Zhuxian Meihua Tu* collected by the Palace Museum. The is a similarity between the technique in the drawing from central panel and Chen's work. The inscription on the central panel may be in different handwriting from Chen's usual signature, and therefore this is possibly not the original signature of the artist.

On the back of the central panel, in semi-cursive scrips are carved six sentences of poems collected from different poets. This was a fashionable form of entertainment for scholars during the period.

The *kehui* technique, which is carving lacquer, is applied on this piece. The craftsman carving the paintings or calligraphy on the lacquer surface which has a similar effect to that of an inscription carved in stone. The varnish lacquer surface provides a black outline for the brownish ash layer beneath which presents an ink rubbing appearance. This type of finish was popular with a small group of the scholars in the Jiangnan area.

008

剔红
山水人物图座屏

明中晚期
高 30 厘米　宽 30.5 厘米　厚 13.5 厘米

剔红座屏，于屏具而言为中小型，于漆器而言属中大型。

屏心双面可观，一面为山水人物图，一大片水域中散布群岛，山势圆浑，造型稚拙，上有亭台、楼阁、树木，仙洞神府所在。近景处一片空地，山岩掩映着殿宇，两士人对坐厅堂，旁有童子手扶隔扇。庭院中一老者挂杖而立，周围有童子持卷轴、杯盘、琴等，台阶下有一童子跪地迎接，似乎暗示来者是位神仙人物。庭院正中有太湖石嶙峋而立，旁生一株古松，蟠绕而上，身姿如龙，直达天际。屏的另一面雕饰牡丹纹，物象平整，茎脉刻画工整细致，枝繁叶茂，布局几乎密不透风。

屏框上方做成大委角，线条柔和，看面起鼓，其上雕饰花卉纹，横向为牡丹、菊花，竖向为梅花，构图奇特，皆是繁密的花叶沿着边框蔓延，中间拱托一朵花头，层次厚密，营造出如同层层相叠的花丛效果。屏框侧面凸起一线，雕为绳状，如同自左、右和上方三面捆拦花枝于屏框上。壶门式披水牙板，牙板为弧形面，较为少见，其上雕水仙四丛，花叶甚繁，依然是草叶丰厚之状。站牙、抱鼓墩处亦雕风格相同的水仙、梅花等图案，唯抱鼓下部件雕刻条带捆扎树叶等物。

此屏的雕漆手法甚为特殊，布局奇诡，山若垒石，人物淳朴可爱，尤其是古松雕法，树干若蛇鳞，松针若轮，针叶皆弯弧状，如同顺时针旋转的车轮。花卉纹装饰则力图表现草木葱茏、丰饶之态，颇具南方风味。相似的例子见于台北故宫博物院所藏宣德款剔红七贤过关人物圆盒（载《和光剔彩：故宫藏漆》件30），另见私人收藏有剔红群仙贺寿图脚踏和剔红山水人物图长方箱。这几件剔红器的图案风格和雕刻手法均很统一，与明清时几类主要剔红风格皆不相符，应是尚未引起关注的某一地域风格。尤其值得一提的是剔红山水人物图长方箱，为对开门，下附底座，其对开门所饰山水人物、立墙所饰凤穿牡丹纹和底座的花卉纹，无论图案还是工艺，均与本品如出一辙，显然是匠师根据同一稿本变化而成。日本出光美术馆藏有一件剔红座屏（蔡和璧《多宝格上的瓷板画》，载台北故宫博物院《故宫文物月刊》第3卷第11期），造型、纹饰与本例近乎相同，唯屏心另嵌有青花瓷板，为庭院仕女人物，为明代空白期风格。其瓷板周围图案与本例相近，只是重新排列安排在瓷板周边。以出光美术馆的这件座屏为参考来断代，本屏的制作时间当在明中期或以后，从图案风格来看，最接近嘉靖时期。

明
剔红镶青花人物故事图座屏
日本出光美术馆藏

Ming Dynasty
Carved cinnabar lacquer table screen inlaid with blue-and white panel
Idemitsu Museum of Arts

105

宋
李迪
《红白芙蓉图》页局部
日本东京国立博物馆藏

Song Dynasty
Li Di
Part of Album of *Red and White Hibiscus Roses*
Tokyo National Museum

Carved cinnabar lacquer table screen

Middle to Late Ming Dynasty
H 30CM W30.5CM D13.5CM

This carving technique is used on this table screen specially creating beautiful fine carved lines to illustrate the tree's trunk in a snake scale pattern. The shape of the pine leaves presented in a circular shape. Similar carving technique may also be seen in the Idemitsu Museum of Arts Japan and the Palace Museum Taipei. These techniques were seen on objects from southern China in the same period.

    The art of carving lacquer is a technique unique to China. In carving lacquer, multiple layers of lacquer are applied onto a substructure in the shape of an object. After which carved to create lush geometric motifs, engaging scenes of figures enjoying social activities, or lively birds flitting among flowers, creating delicately carved backgrounds in which different geometric designs are used to show earth, water, and sky.

009

黑漆框
剔红云龙纹座屏

清早期
高 36 厘米　宽 37.5 厘米　厚 15 厘米

　　屏心正面为剔红云龙纹，二龙一升一降，争夺火珠，下有海水江崖。背面黑漆地上百宝嵌折枝花卉，以螺钿、染牙、玳瑁、寿山石、绿松石等嵌成，分别为菊花、佛手和枝叶，寓多寿多福之意。

　　屏框及屏座髹黑漆，漆灰坚硬，漆层薄而致密，几不见断纹，发色纯正，形成浑然一体的效果。屏框较宽，内外圆角，柔和而文雅。站牙、抱鼓、屏墩横向宽厚而纵向短促，形成厚拙可爱的效果，有建筑意趣，与常见座屏比例迥异。其抱鼓较大，夹挟屏框，有宋代座屏意趣。

　　故宫博物院收藏有同款座屏，唯尺寸稍小，高20.6厘米、宽23.2厘米、厚11厘米，正面镶嵌剔红花鸟图屏心，背面百宝嵌与此屏相同。其黑漆框上有暗描缠枝莲纹，为康熙时期特征。两者均是先有剔红屏心，再制框、座。故宫所藏者屏心为典型明中期剔红制品。此件屏心从龙纹特征看应为清早期制品。两者的尺寸差异，亦是因屏心大小不同而致。黑漆和百宝嵌的做法，是扬州等地区工艺特征。

明
仇英
《赵飞燕外传》卷局部
私人藏

Ming Dynasty
Qiu Ying
Part of Handscroll of *Anecdotes of Zhao Feiyan*
The Private Collection

Black lacquer framed carved cinnabar lacquer table screen

Early Qing Dynasty
H 36cm W 37.5cm D 15cm

On the front of the central panel is carved two dragons attempting to catch a fire ball with one dragon upwards and the other dragon downwards with a sea and mountains pattern providing the background.

The rear of the central panel is inlaid with colourful precious and semi-precious stones and materials which include lapis lazuli, turquoise, mother-of-pearl, rose quartz, malachite and amber. This technique is known as the 'Hundred Treasure Inlay (*Baibao Qian*)'. The artist's refined level of skill and expertise in working with the various materials is displayed in the inlay method, a bolection-like technique in which the inlay projects beyond its frame and thus creates a surface of different levels with a strong three-dimensional quality.

Black lacquer technique on this piece, is finished with a dense and smooth surface look and negligible cracks. The wide frame and boarder and stocky foot has particularly elegance body features.

This is similar to a piece in the collection of the Palace Museum. The front of the central panel is carved illustrating brids and flowers. The back is inlaid with colourful precious stones. Both sides of the screen have used carved cinnabar lacquer on the central panel. From the technique used here, we may deduce that the artist would have made the carved lacquer panel first then worked on the frame and foot determined by the size of the central panel. From the carving details of the dragons on this piece it may indicate the date of this piece is from the early Qing Dynasty. The lacquer technique and inlaid precious stone technique may be from Yangzhou.

清
黑漆框剔红花鸟图座屏
故宫博物院藏

Qing Dynasty
Black lacquer framed carved cinnabar lacquer table screen with flower and bird pattern
The Palace Museum

010

紫漆百宝嵌
秋郊饮马图座屏

明末清初

高 21.3 厘米　宽 26 厘米　厚 12.5 厘米

中小型横幅座屏，砚屏之属。

屏本是横幅，却在屏心框下又设一框，中有间隔如开光，使得屏心更加狭长，恰如一幅长卷展开，更觉雅致。这种处理手法宋时已有，甚至有将竖屏通过镶装绦环板等方式，使得屏心狭长如长卷一般的做法。

屏心两面可观，皆有黑漆子框。正面为秋郊饮马图，红漆为地，大片留白为江河、天空，近处嵌成坡岸（嵌物已失）。岸上一株秋树，树叶以螺钿嵌成，树干堆漆而成。树下马以整块螺钿雕嵌而成，皮壳温润如玉，眼球点以黑漆，深色材料镶嵌马蹄。马低首欲饮，前蹄微抬，马尾卷自左后蹄内侧缠卷，似是不经意地甩动而致，鬃毛、马尾等处以细笔划出毛发。画面左上角露出远山一角，以玳瑁、螺钿、寿山石、虬角嵌成。此图布局空灵，红漆艳而不俗，螺钿嵌成的马和树叶在其衬托下甚为醒目，俨然一幅重彩山水画，使人第一眼就想到了赵孟頫《秋郊饮马图》。历来所见百宝嵌图案，多以丰富的镶嵌物象为表现对象，此处只见饮马，余则便是宽阔宁静的河流被落日染红。今日视之，数百年悠悠而过，长河不尽，饮马恒在，却蕴涵着时空在静止与变化间转换的玄机，触动着观者的思绪。

背面黑漆为地，嵌饰稻禾、菊花，布局空灵，嵌饰精巧雅致，唯意境不如正面深邃。

屏框紫漆，壶门牙板，坐墩和站牙轮廓皆作连弧状，直露圭角，类布币，有金石意趣，是明末清初时仿古之风的体现。

座屏整体薄灰薄漆，漆质细腻坚硬，几不见断纹，从嵌饰手法和图案风格来看，接近扬州地区风格。

Purple lacquer inlaid precious stone table screen

Late Ming to Early Qing Dynasty
H 21.3cm W 26cm D 12.5cm

Regular horizontal shaped table screen. The open space between the central panel and frame ccentuate the themes depicted on the central panel. A white horse is drinking water from under the tree. Of all animals, the horse has been represented in the widest range of contexts, reflecting this animal's central role in Chinese history, literature, and mythology. In the early Yuan period, when the ruling Mongols curtained the employment of Chinese scholar-officials, the theme of the groom and horse-one associated with the legendary figure of Bole. Bole's ability to judge horses had become a metaphor for the recruitment of able government officials became a symbolic plea for the proper use of scholarly talent. White, a well-fed hunk of a horse with a powerfully muscular chest, beautifully stylized with charmingly caricatured details, is one of the many attractions of 'power and virtue'. The white horse has particular power and associated with Buddhism mythology power in different ways.

The lacquer technique on this piece could be from Yangzhou area, smoothly finished lacquer without cracks.

齐白石
《日行千里》轴
蔗园藏

Qi Baishi

Hangingscroll of *Cover a Thousand Li in a Single Day*

The Zhe-garden

011

紫檀嵌石、牙
二甲传胪图插屏

清乾隆
高 31.8 厘米　宽 24.3 厘米　厚 14 厘米

屏框立帮式，宛如方盘竖立，形成如舞台般的空间，内镶嵌图案为饰。屏心嵌石、牙为二甲传胪图。画面左上方，五株芦苇，穗头丰硕，以石雕嵌而成，秆、叶以染牙嵌成，作翠绿色。画面中下部，以石嵌成两只张牙舞爪的螃蟹，以蟹螯钳夹穗头，谐音"二甲传胪"。石黄色，质近祁阳石。

古代科举考试，二甲为殿试第二等，赐进士出身，传胪为唱名传呼之意，始于宋代。金榜题名是人生重大幸事，故古人以螃蟹和芦穗谐音来讨彩头，亦有出现一只螃蟹或三只螃蟹的一甲传胪、三甲传胪图案。螃蟹芦穗的组合图案，至迟宋画中已见，当时为写生小品，明清时期，开始具有吉祥寓意。

清宫旧藏清早期黄花梨独占鳌头图高盆架，其中牌子绦环板上有一甲传胪图案；另有清中晚期楠木镶祁阳石三甲传胪图插屏。这两件家具虽然可判断为地方进贡之物，但无论接受者还是进贡者，均未太在意皇帝无须金榜题名的事实。此亦提醒读者，中国古代纹饰的应用，虽有很多具体的寓意，但有时仅是图案化的纹饰，只要不是僭越，则取其吉祥，并不拘泥于具体所指。

屏座立柱亦随屏框而加宽。变体石榴头式站牙，其与绦环板、披水牙板上皆浮雕勾云纹，为清乾隆宫廷器具中十分经典的仿古纹饰。长方形屏墩，下挖亮脚，看面剔地浮雕打洼拐子纹。

此屏出自宫廷，芦穗螃蟹图案为其增添几分野逸气氛，搁置案头，大有湖畔水滨之感。

明
沈周
《写生册》局部
台北故宫博物院藏

Ming Dynasty
Shen Zhou
Part of Album *of Paint from Life*
The Palace Museum, Taipei

Red sandalwood inlaid ivory and stone table screen

Qianlong Period, Qing Dynasty
H 31.8CM W 24.3CM D 14CM

The combination of reed and crab in Chinese art commenced from Song Dynasty. When this combination was used in the Ming and Qing dynasties, it has auspicious meaning.

There are two pieces of imperial furniture collected by the Palace Museum which have employed same topic. It may be as a tribute gifts from local governors to the emperor. The original idea was that these symbols would bring good wishes for the young scholars joining the national exam to become officers. When this topic was used by the court, it has auspicious meaning.

012

黑红漆麒麟纹座人物故事图插屏

元

高40厘米　宽34厘米　厚26.5厘米

屏心为髹漆木板一块，不设边框，即所谓的"插牌子"式。屏心上转角作委角，其上髹黑漆，略泛紫，正面原有彩绘图案，勉可见为人物故事图，右下方可辨者为一微髭男子，戴幞头、着长袍、手持扇，其他各处影影绰绰可见数个人物，亦持扇，两边有两个红衣窄长袖的随从状人员，右上角似为宫殿。背面光素无纹。黑漆质地甚坚，发断细碎如冰裂，有零星剥落，如星空之灿。

屏座立柱上雕仰俯莲纹，花板尖而高凸，收腰较细，形态精巧。并置双绦环板，髹红漆，上有剑环式开光，内透雕麒麟纹，双面工，图案原有描金，掉落较多，背面保留稍多。麒麟为相向的跪卧状，形若羔羊，尖喙凸睛，须发自角旁凝为一束前伸，颔下有须，身上鳞片宛然，尾高高翘起，较大，三瓣花叶状，有飘带状卷草自前腋下缠绕后蔓延，若火焰翼。麒麟纹造型简练，但无一不备，形态生动，憨态可掬。

站牙髹红漆，原有描金镂空花纹，惜已残失。壸门式披水牙板，两头有相抵的卷草叶装饰，亦髹红漆。屏墩抱鼓式，下方衍为卷叶状回卷，一叶片变为卷珠，另一为舒展的花叶，垫在扁圆状鼓下，叶片看面上方起脊，加以致密的漆表，质若铁石。屏墩下设圭角，上起一线，下方四面膨出，斜向铲出壸门线。

此屏出自山西，漆质保存相对完整，与山西家具常见的雄浑拙朴风貌不同，精巧细腻，为同类中之佳品。

经碳十四检测漆片，为1282年±25年，可资断代参考。这与笔者起初断代为明中晚期相差较多，亦说明有些家具的断代需要重新审视，虽然大部分情况下断代偏早，但是也确实存在偏晚的情况。回头来看此屏，其瘦劲的柱头莲花纹、抱鼓下遒劲的卷叶和古拙的麒麟纹，确有一些早期特征。此屏的科技检测结果，令人惊喜，明中期以前的传世家具实在少见，这无疑是重要的一例。

Black and cinnabar lacquer table screen

Yuan Dynasty
H 40CM W 34CM D 26.5CM

This is a surprisingly rare discovery of a Yuan Dynasty table screen. The central panel is formed by a single wooden plate with concave corners. The black lacquer on top of the wooden body has a slightly purplish hue. The front of the central panel may have included a coloured lacquer drawing which has been lost through the passing of time. The sole remaining evidence appears on the lower right corner illustrating a scholar holding a fan, there are two servants behind the scholar, the building seems to be shown in the upper right-hand corner.

There is no trace of drawings on the back of the central panel. The quality of the black lacquer has a moderately solid feature with very fine cracks, some of which have become loose and lost from this piece.

The lotus flowers carved on the stand are a pair of *Qilin* inside of the carved panels which are directly below the central panel. *Qilin* open carved technique, including remaining traces on the trunk of the *Qilin* showing it may originally have used the *miaojin* technique.

This table screen has its origin in the Shanxi province. The lacquer layer on top of the wooden body is in good condition. Carbon-14 dating of fragments of lacquer from this piece reveal it is dated around 1282 ± 25 CE.

013

红漆撒螺钿框
山水图座屏

清早中期
高 22.8 厘米　宽 20.5 厘米　厚 11.5 厘米

　　中型座屏，两面可观，正面黄漆地上墨笔绘山水图案，用笔轻快，寥寥几笔，勾勒亭台楼阁、卧波小桥、山岩瀑布，留白较多，意境简况。背面黑漆地上，饰博古纹，为花瓶、牡丹、磬、水盂、古鼎之类，物象以黑漆描绘轮廓，红漆作底，金漆渲染。

　　屏框平素，红漆地上撒螺钿为饰。下设绦环板，有笔管式开光。披水牙板，牙头处做成刀牙状。连弧状站牙。拱桥式屏墩，上沿拱起较多，两端下方收进，内涵力量。

　　此屏的撒螺钿和黄漆地都有很浓厚的装饰意趣，红漆发色艳而不俗，陈设案头，颇增亮色。其连弧状站牙等造型和漆质特征，与本书所录紫漆百宝嵌饮马图座屏相近，同属扬州地区风格。

Cinnabar lacquer with crushed mother-of-pearl framed table screen

Early to Middle Qing Dynasty
H 22.8cm W 20.5cm D11.5cm

Medium size table screen. The central panel illustrates a mountain and water scene on top of a yellowish lacquer background. The back of the central panel is painted a group of antiquities, including ancient bronze vessels, water pot and flowers reflecting the taste of the literati at the time. These bronze vases used cinnabar lacquer on top of black lacquer finished with golden lacquer showing the beautiful colour contrast.

The brushed mother-of-pearl frame and yellowish lacquer shows special decoration characters. The designs and techniques employed on this piece indicate that its origins are in the Yangzhou region.

014

红木框
黑漆彩绘博古图座屏

清中晚期

高 31.5 厘米　宽 34.5 厘米　厚 15.8 厘米

横屏式，因下设绦环板，屏心更狭长，如长卷展开。屏心黑漆彩绘博古图，右上角有半展的卷轴，上绘兰花，题字为"啸傲烟霞""过眼诗书雪亮，束身名教风流"，前者出自元张可久《折桂令·村庵即事》"掩柴门啸傲烟霞，隐隐林峦，小小仙家"句，后者为清代名联，原作"到眼诗书皆雪亮，束身名教自风流"。画卷后半掩梅花、琴书，落半月印"乐琴书"，出自陶渊明《归去来兮辞》"乐琴书以消忧"。画面前方设卷几，造型曲折，上陈鼎、书，旁置水盂、瓶等。左面双螭耳瓶中插玉兰花，瓶身题"守口如"，即守口如瓶之意。瓶旁绘印张二，上为白文"楚乔"，下为朱文"黎培树印"。此处博古图绘制随意，有民俗趣味，寓意则多有逍遥自得之意。画面上方横题"静观自得"，出自宋儒程颢《秋日偶成》诗"万物静观皆自得，四时佳兴与人同"句。

屏背红漆隶书几句《兰亭序》，字边皆勾金边，右上角椭圆形朱文印"吟风弄月"，左下落款"竹村王采亭"，朱文印"竹村""王采亭印"。

红木为框、座。宽边框，平素无纹，下装绦环板，开鱼门洞，壸门式披水牙板，站牙厚硕，底座拱桥式，均保持简素风格。其造型明风浓郁，落落大方，格调较屏心为高。

红木所制家具，多自乾隆时始，此屏综合屏心判断，应为制于清代中晚期之物。

Hongmu framed black lacquer polychrome table screen

MIDDLE TO LATE QING DYNASTY
H 31.5CM W 34.5CM D 15.8CM

Table screen in a horizontal rectangular shape with carved panel below central panel. The whole frame has stocky features without any enhanced decoration pattern relying entirely on the beautiful wood grains as embellishment. The central panel depicts antiques or *bogu* scene. The orchid illustration is on half opened handscroll on the top right corner. The water pots, books and plum flowers presented in a casual environment showing folklore characters.

The back of the central panel citing *Orchid Pavilion Gathering* is in official script, the gold lacquer outline technique used on characters.

The *hongmu* furniture commenced from the middle Qing Dynasty. The use of this material indicates this piece may be from the mid to late Qing Dynasty.

015

黑红漆
山字式可折叠座屏

明晚期

高 73.3 厘米　通宽 109 厘米　厚 35.5 厘米

屏三扇，中高旁低，山字式。两侧为可折叠式屏扇，内侧边框以铰链与中扇相连，外侧边框下方出头落地为支点，其上框与中扇上方的绦环板下沿相齐，宽为中扇之半，合起来如关门状。

中扇座屏式，设一周绦环板，屏心下方增设两格绦环板，使屏心位置升高。绦环板皆镂空斜卍字纹并髹红漆装饰，除对称部位花纹相同外，又各有方向、大小、形态变化。两边扇亦设一周具各种变化的斜卍字纹绦环板，上方绦环板为六边形套卍字纹或六角花纹，如同窗棂。边扇屏心狭长，如一副对联。屏框看面皆起剑脊棱。屏心素黑漆，漆质细腻，断如鱼鳞。

这种山字式可折叠座屏并不多见，中型如此例者见有几例，另见有大型落地者，多以黑漆为主，上有红漆或彩绘修饰，均为典型山西家具特征。度其用法，或以纸张作书画，两侧为联，贴于其上。两边扇平时关合，屏心内容与时令或使用场合有关，或为神佛、圣贤、祖宗等，逢祷告、祭祀时方展开使用。

山字式座屏在魏晋时已有雏形，为一种三面围合的屏，至迟五代时期已基本定型，如王齐翰《勘书图》、周文矩《重屏会棋图》中皆有所见。陆游有"水纹藤坐榻，山字素屏风"诗句，显然在宋时已甚为普遍。这种三面围合的屏风，在宋时有名为枕屏者，可围榻一周，朱熹《家礼》记冬至祭祀，"屏风如枕屏之制，足以围席三面"，亦是此类。今日所见山字屏多为清代宫廷所用的固定式大座屏，这种两旁连以铰链可活动的实例甚是少见。

Black lacquer folding table screen with geometric lattice open work

Late Ming Dynasty
H 73.3CM W 109CM D 35.5CM

Mountain shapes folding table screen, the central panel being taller than the two side panels resembles a mountain shape, the width of the two side panels being equal to the width of the central panel when closed. When side panels are folded in, the scene appears as a door of a Chinese house. Metal hinges are used to attach the central panel to the side panels. Carved panels made of short members joined together to form the *wan* pattern, a popular declarative motif of the Ming period. On this piece, there are seven different *wan* patterns used in different areas presenting rich beautiful patterns. The central panel of each screen is finished in black lacquer, the lacquer having a smooth appearance with fine cracks resembling fish scales.

This is a rare piece. A three-panel folding screen of small table dimensions. The size of this piece indicates it may have been used for ritual purposes, such as ancestral shrine or a divinity shrine. Usually, three panel folding screens are large size furniture pieces from the Shanxi region, employing the black lacquer or cinnabar lacquer or polychrome lacquer techniques. The use of this type of screen in China may have commenced during the Wei and Jin dynasties, used by high-ranking officers or aristocratic families in seating area for causal and formal events.

山水寄情
LANDSCAPE SUSTAINS ASPIRATION

016

黄花梨
镶大理石座屏

明晚期
高 23.8 厘米　宽 24.7 厘米　厚 12.2 厘米

　　典型砚屏，高、宽相若而近方，但下方留空档，装壶门券口牙板，屏心位于上部，长横幅状。

　　屏心大理石，两面可观，一面纹若奔兽，仰头甩尾，自画面中下部驰向右上部；又若一块嶙峋的横峰赏石。另一面若苍山落雪、水面凝冰。

　　屏框方直，两边竖框与横框格角相交后变窄，直落于托泥上。屏框下方空档装壶门券口牙板，曲线柔和，竖牙条至底端为勾卷花叶状。托泥为"亚"字形，中间落膛镶装绿石，可盛文房杂物；两端承接屏框、竖牙条和壶瓶式站牙。

　　明清座屏样式之成型，当自宋始。至迟在北宋晚期，已经出现屏心位于中上部为横幅，下方留空档镶装绦环板或牙板的座屏，空档高度达整个座屏三分之一以上。这种制式的座屏即使在明清绘画中亦为常见，然传世实例，无论大小、材质，屈指可数。

　　宋时砚屏图像少见，实例阙如。张择端《清明上河图》(故宫博物院藏) 近末段 "久住王员外家" 右侧楼上，绘一个背靠书法座屏的读书人，其书案上就摆着一个正对他的小砚屏，造型近方，屏心位于中上部，下方似为绦环板或牙板，与本例造型相类。

　　砚屏之制始于北宋，南宋赵希鹄《洞天清录》"砚屏辨"："古无砚屏。或铭砚，多镌于砚之底与侧，自东坡、山谷始作砚屏，既勒铭于砚，又刻于屏，以表而出之……屏之式，止须连腔脚高尺一二寸许，阔尺五六寸许，方与盖小砚相称，若高大非所宜，其腔宜用黑漆或乌木，不宜用钿花、犀牛之类……"

　　日本德川美术馆藏有一件明清之际的黑漆百宝嵌座屏，高22.6厘米、宽18.6厘米、厚8.8厘米，上为屏心，正面嵌成松鹿图，背面嵌萱草，下有壶门券口，与本例甚为接近。明万历朱守城墓出土的紫檀镶大理石座屏亦与此例相近，但其前方另附一体的条桌状笔架，更为复杂。美国波士顿美术馆"屏居佳器"展亦见与之相似的一件黄花梨镶大理石座屏。托泥镶嵌石板者，目前仅此一例。当然，在强调本例罕见的同时，还请读者更关注于其优美的造型和石片。

黃花梨鑲大理石座屏

許叔重說文云屏者蔽也有內斂之美大隱之質研屏之式或肇於宋趙希鵠洞天清錄載古有研屏或銘硯多鐫於硯之底興側自東坡山谷始作研屏既勒於研又刻於屏以表而出之云玄觀是屏高寬相仿下飾以壺門券口牙板取勢溫潤柔和上嵌雲石一方作雙面之景觀其一紋若狡猊隱現雲中顧盼之姿呼之欲出其二仿佛雪景深遠得徽廟雪江歸棹之境屏研相宜自是文房雅陳之樂更於案頭增一分遠山悠揚之意
己亥年冬月為蕉園藏屏題金陵劉丹

刘丹
跋 "黄花梨镶大理石座屏"
2019

Liu Dan
Postscript of 'Huanghuali and dali marble table screen'
2019

HUANGHUALI AND *DALI* MARBLE TABLE SCREEN

LATE MING DYNASTY
H 23.8CM W 24.7CM D 12.2CM

Rectangular cream to beige *dali* marble panel with white, grey and brown inclusions, mounted in a *huanghuali* frame above a cusped spandrelled apron.

Table screens of this size are called *yanping* (inkstone screen). They were used for shielding ground ink powder from the wind. Inkstone screens are delicate table-top additions to a scholar's implements of brush, inkpot, and inksticks which are little known or recognised.

017

乌木
镶大理石座屏

明末清初
高 43 厘米　宽 39 厘米　厚 15.8 厘米

　　屏心大理石，米家山水意味。黛色山峦两重，每重又各化为两重相叠状，远山缥缈而舒扬，近山凝重而沧桑。余处白色为云、天，年久泛黄，皮壳深浅不一，烟云弥漫。

　　乌木为框，形简而端庄，有雕塑感。框看面打洼，上方转角做成委角。屏下设素刀牙板，牙头短小质朴。站牙为桨腿式，曲线柔和而微妙。屏墩略呈如意状，亮脚处垂注膛肚，两足内卷而翘，曲线柔和而内蕴力量，若猛兽曲爪。

　　屏的比例处理甚好，观之近方，实是纵大于横，但加以屏墩抬高，兼上方委角，形成方正温和的视觉效果。

　　乌木是一种珍稀硬木，明晚期时开始制作家具、器具，然以之制屏的记载宋代已有。南宋赵希鹄《洞天清录》"砚屏辨"中载有以蜀中松林石制砚屏者，"其腔宜用黑漆或乌木，不宜用钿花、犀牛之类"。"腔"指屏框，"腔脚"即屏墩，赵希鹄所述者为小砚屏，以黑漆或乌木制者，概取其文雅素朴。

　　石纹山势作一重或两重，石色白多于黑的云山式大理石，多见于一些风格古朴的座屏等镶嵌家具。如明万历朱守城墓出土紫檀镶大理石座屏，石板纹路风格即与本器相近。推想或是明人选石特点之一，然是否确论，需更多论证。屏的桨腿和刀牙板样式亦为晚明特征，保守判断，此屏当制于清初，不排除明代制器的可能。

WUMU AND *DALI* MARBLE TABLE SCREEN

LATE MING TO EARLY QING DYNASTY
H 43CM W 39CM D 15.8CM

The creamy-white marble with brown and dark greys, forming a natural design of layered mountainous boulders, displayed in a *wumu* frame.

The marble in this piece appears as a mist surrounding the group of mountains encouraging the viewer to recall ancient landscape paintings in China. These landscape-like streaks may refer to the Ming people's aesthetics on selected marble.

Wumu or ebony, has a fine closed grain which is very brittle, and the colour may be pure black to black-brown. Because it grows as a small-diameter tree, it is rarely used as primary material for large pieces of furniture, but more often shaped into secondary decorative elements or as small precious objects.

This piece is designed in perfect proportion, displaying a square shape rather than the usual rectangular. The secret is in the feet of H-form trestle base in stocky feature and top two incurvate corners presented a smooth outline appearance.

018

黄花梨
镶绿石座屏

明末清初
高 35.8 厘米　宽 32 厘米　厚 16 厘米

　　中小型座屏，比例方正，属造型最简练的一种。

　　屏心为黄绿色石，依纹路稍加磨制成瑞兽观日（或月）状，瑞兽尾高翘，略备四足，登高回首观看，圆日当空。物象形态概括，影影绰绰，若水中倒影，弥漫四周的纹路若云、雾、星空，不可尽识。此类绿石产地不明，有虢州说、祁阳说等。绿石磨制为瑞兽日月状，亦属较经典的做法。背面石片较抽象，无具体形象，为星河、云浪状。

　　屏框宽阔，与屏心对比明显，使观者自然地聚焦于屏心。黄花梨顺纹料，甚为含蓄，原皮壳，色如烙饼，深浅变化自然。攒框后竖框下方伸出，榫接于桥梁式屏墩上。屏下设刀牙板，牙头短阔。桨腿式站牙。

　　此屏皮壳状态更近北方，绿石磨制瑞兽等图案的做法，常见于山西地区家具。其造型简朴，明风甚浓。

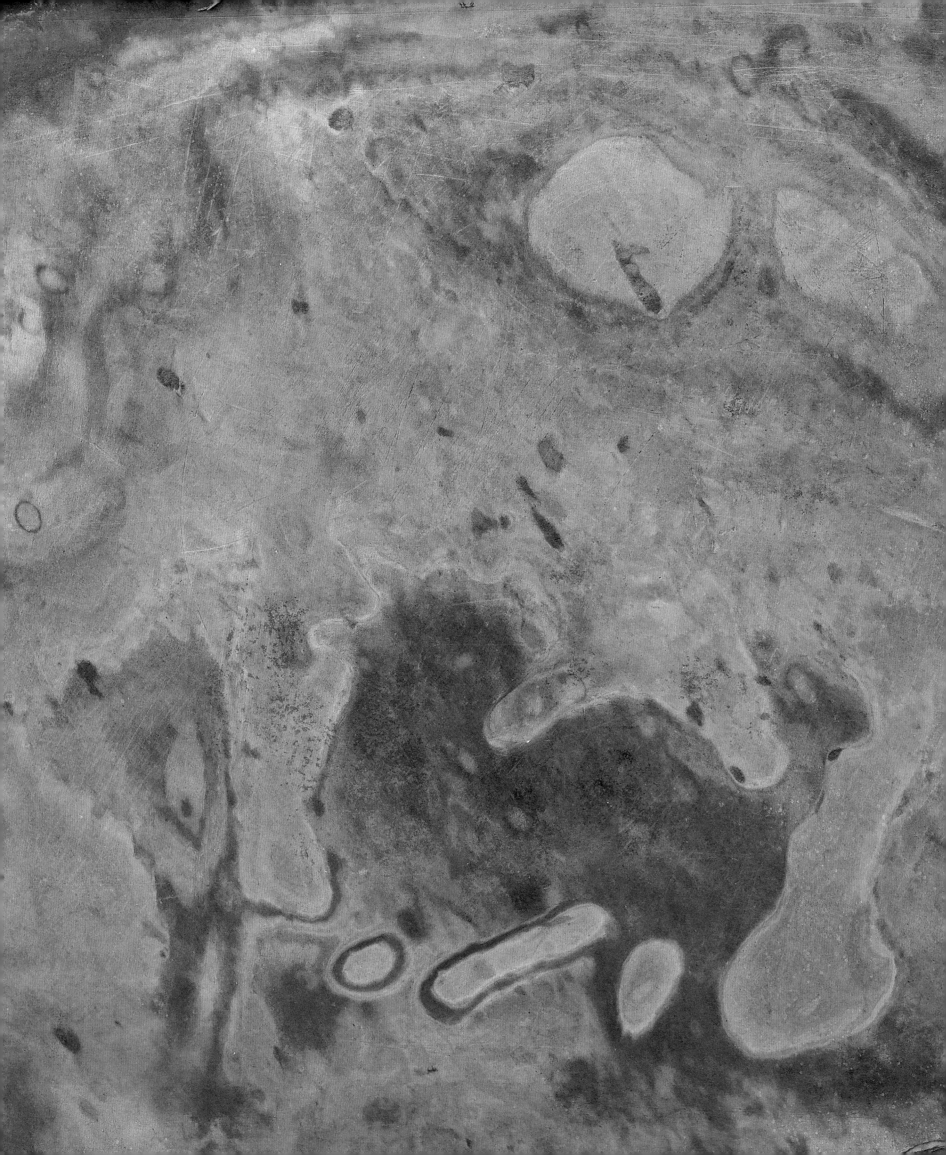

Huanghuali and green stone table screen

LATE MING TO EARLY QING DYNASTY
H 35.8CM W 32CM D 16CM

The outstanding qualities of green stone was its fine and smooth textures and as such was valued as the best material for inkstone since the Tang Dynasty (618-907CE).

The beautiful stone grain invokes mysterious animals looking at the sun or the moon from the back with the clouds or the mists floating between the animals and the heavens.

The *huanghuali* frame with wooden grains devoid of caving or inlay decoration allows the beautiful colour of the green stone and natural grains to shine without competing with its surrounding frame.

019

黄花梨
镶白石座屏

清早中期
高 43.5 厘米　宽 49 厘米　厚 19 厘米

屏心白石质洁如玉，皮壳层次丰富，如水波、风沙，光泽内蕴，其上淡淡几重灰色山脉，只见轮廓一线，若有若无如水渍，是以刀凿刻成。这种在石板上錾刻纹路的做法，时有所见，本书所录就有两件，对制者的要求较高，达到"虽由人作，宛自天开"方为佳。

屏框、座黄花梨制，深褐色原皮壳，更加映衬屏心。造型简约，下附牙板，中间雕变体如意云头，两端饰拐子纹，轮廓波折起伏较多。变体宝瓶式站牙，拱桥式屏墩。此屏所用为黄花梨中小料，多枝节，甚致密，上布有节疤，如水之荡漾，与屏心浅淡的纹路有相映之趣。

牙板背面不再起线，显示前后有别，但今日背面依然可观，屏心皮壳色深，有苍古味，边框皮壳老辣，甚合好古者之癖。

Huanghuali and *baishi* table screen

EARLY TO MIDDLE QING DYNASTY
H 43.5CM W 49CM D 19CM

The *baishi* has a clear and smooth texture similar to jade. The rich layers look like waves of the sea or clouds. The outlines caved on the top of the panel appear as water marks, being a rare decoration technique used on the table screen (another two pieces in this collection also use this caving technique).

The frame and stand are made of *huanghuali*. The dark brownish frame is the prefect colour choice for this stone panel. The *huanghuali* material used here is predominately small branches accentuating the grains as a natural decorative pattern for the frame. The apron is decorated with a carved cloud and angular hook pattern.

020

黑漆框
镶大理石座屏

明晚期

高 69.5 厘米　宽 66.5 厘米　厚 33 厘米

　　座屏中型偏大，面心近方，镶嵌大理石。石板两面皮壳各不相同，一面老辣苍古，烟熏黄色，山峰一座雄峙，空处质若风沙飞舞；另一面白地上青灰色纹路，若风起云涌，气象甚佳。

　　黑漆为框，质纯色正，上有蛇腹断，形甚简练。下附壸门式披水牙板，两端有对卷云纹。壶瓶式站牙，形饱满，亦饰对卷云纹。屏墩抱鼓墩，制式规矩而简练。下方附"亚"字形地栿。

　　此屏从站牙、抱鼓墩等造型来看，与本书收录几件明晚期座屏相类，信为相近时期制品。其边框与底座比例拿捏甚好，简练古朴，得典雅之美。

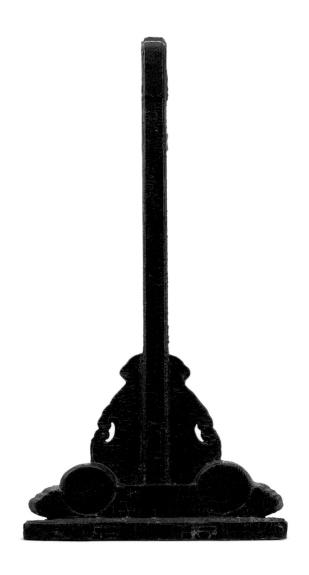

Black lacquer framed *dali* marble table screen

Late Ming Dynasty
H 69.5CM W 66.5CM D 33CM

This is a mid-size screen with marble panel close to a square shape. The one piece of marble displays different colours on each side which makes this a rare and interesting piece. One side had smoky brownish colour similar to an ancient painting whilst the reverse side is of a white greyish colour featuring black abstract waves.

The pure black lacquer frame designed in the simplicity style and has snake scale cracks. Outward sloping scrolled apron on an H-shaped trestle stand with cloud patterns and upright brackets also having cloud decoration patterns.

The design of this piece has similar features to other pieces in this book which referred to the late Ming Dynasty table screen design, therefore dating this piece may be of similar antiquity.

021

紫檀
镶大理石座屏

明末清初

高 34.5 厘米　宽 36.5 厘米　厚 18.5 厘米

　　中小型座屏，宽边，小屏心，紫黑色紫檀与白色大理石纹相映衬，甚为醒目。

　　屏心近方，正面白质黑章，中间耸立一峰，较巍峨，上有丫杈斜伸，旁辅小山峰，上亦有丫杈状石纹，余则为地，云水一色，皮壳层次丰富。背面本无纹路，皮壳如陈年老纸，亦可充一时之赏玩。

　　石板山峰上有丫杈状纹，属较程式化的纹路。概古人取石，相自然纹路，稍加磨制整理，略成几类较为常见的图式。大理石多见山峦、树木纹，绿石则多见猫虎瑞兽、人物仙佛等纹路。以出于自然、因势而成为妙。

　　屏心石板与屏框结合已较为松散自然，可见其结构为石板边缘磨薄为榫舌，甚短，堪可供边框管束即可。

　　座屏的边框处理，常见有两种，一种四面边框同宽，下附底座；另一种下框变窄，加以其下的横枨、牙板宽度，方与其他三框宽度相近（有时稍宽些），取得视觉的均衡。这两种方式，前一种注重屏框的交圈，后一种注重整体和谐。此屏即采用后一种方式者。

　　屏下牙板，剔地沿边起平阳线，两头和中间各垂下为回纹。屏墩与屏框间设连弧状斜枨。屏墩平素硬朗，质如铁石，拱桥式，外转角有连弧状挖缺造型，内转角处亦做成连弧状凸起以呼应。

　　牙板下垂的回纹和屏墩下方平齐，已可触地，按照常规的造型规律，应该在下方留有一定空间，但审查牙板、屏墩，原制如此，究其原因，或座屏下还有可拆卸的地栿，或设计即如此，牙板可提供少许力学支持，尚待更进一步研究。本书另收录一件黄花梨夔纹框镶大理石座屏，亦是如此。

　　连弧状斜枨的做法起源甚早，从座屏发展历史看，起初只设屏墩，后渐出现屏墩与屏框间设连弧状斜枨的做法，至迟五代时已有，北宋时大量应用，而站牙的使用要晚一些，约在北宋晚期，其后多见站牙而少见连弧状斜枨。此外，还见有一些连弧状斜枨和站牙并设的做法，一般多见于金元风格的屏具中。

　　此屏造型古朴，牙板处回纹的做法与清代流行的回纹样式也不尽相同，制作当在明晚期至清中期这一区间。

　　此屏屏心石板边角稍有崩坏，上方已脱槽而出。边框榫卯数处已有开缝，木料亦有几处鼠啮痕迹。需特别强调的是，这种松松垮垮的状态，若不是到了影响其牢固、加速毁坏的程度，以不修理为上。此屏从结构来看，再放几百年依然无虞；从审美来看，松散自然，整体浑然一体，若是补全残损，合紧榫卯，反而失于自然，趋于僵硬，为添足之举。近些年来，木器家具皮壳渐受重视，大家开始注意保护，但修复问题尚在讨论，以笔者之见，若非影响寿命之结构问题，皆可不修，尽量保持其原始信息为最佳的保护措施。此屏状态，正是佳例。

Red sandalwood and *dali* marble table screen

Late Ming to Early Qing Dynasty
H 34.5cm W 36.5cm D 18.5cm

Middle to small size table screen, comprising a small marble panel surrounded by a wider frame. Beautiful colour contrast between red sandalwood and white-greyish marble.

The shape of the marble panel is square, one side of panel contains ink-coloured grains, appearing as an ancient landscape painting, with a rock under the twisted tree trunk, in front of running water, clouds floating in the sky. All of these abstract images are created from smoky grain layers in the marble.

The craftsman would have selected the marble for its natural grains then follow the grains by polishing the marble presenting images of interesting scenes or mysterious creatures.

The corners of marble panel has been damaged, the top is out of the frame, as result the joins of the frame have minor splits. There are minor surface scratches throughout.

The marble panel grounded thin forming a tongue for joining to the frame. The specific characters of the frame, the depth of lower frame compared to the depth of rail and apron are almost equal to the sides and top of the frame.

The apron below the frame had relief carved line, with rectangular spiral patterns decorated in the middle and sides, which reaches to the ground. In general, there are gaps between the apron and the ground. Through examination we have confirmed that this is the original production. Its particular design requires further research. A *huanghuali* frame carved with dragon decoration pattern and *dali* marble table screen in this book has similar features. The techniques used on rectangular spiral are close to late Ming to middle Qing Dynasty.

022

黄花梨
镶大理石插屏

清早期
高 40.8 厘米　宽 38 厘米　厚 18.5 厘米

屏心大理石，黄地，棕褐色纹，变化奇诡，若火焰流动，皮壳厚重老辣。两面可观，正面若黄河怒涛，又若沙漠风暴，雄浑苍劲，气势骇人。背面气象迥异，如风止波静，只是尘埃未定，尚弥漫于空，远处隐约见一峰耸立。两面石纹如画，亦中亦西，甚耐观想。

黄花梨框、座，深红褐色原皮壳，造型方正规矩，为插屏中最基本之样式。刀牙板，牙头阔而方正。素立柱，桨腿式站牙较宽阔，拱桥式屏墩，下方亮脚较矮。整体较朴拙，与屏心意趣相合，追求厚重的效果。

HUANGHUALI AND *DALI* MARBLE TABLE SCREEN

EARLY QING DYNASTY
H 40.8CM　W 38CM　D 18.5CM

Huanghuali frame and stand has reddish-brown colour in rectangular shape format in a standard way. Both sides of *dali* marble has abstract grains, the dark brown grains on top of yellowish background, appears as powerful moving flames, or storm, the grain on the other side imagines a mountain covered by mist. The apron and stand are unembellished with clean lines and functional design.

023

黄花梨
镶大理石插屏

清早中期
高 48.5 厘米 宽 43 厘米 厚 19.5 厘米

屏心大理石近方，两面可观，地子洁白如玉，其上有灰、黄色纹路，斜向一道山脉，灰色纹路中间杂绿色，在白色地子衬托下甚觉清新，余则淡色纹路如云似雾，弥漫周匝。

黄花梨素屏框，宽窄适度。屏座造型甚简，刀牙板较宽阔，桨腿式站牙，下方收进。拱桥式屏墩，下挖矮亮脚。

HUANGHUALI AND DALI MARBLE TABLE SCREEN

EARLY TO MIDDLE QING DYNASTY
H 48.5CM W 43CM D 19.5CM

White grey *dali* marble panel is almost square shape, has beautiful gray and greenish grains on both sides. The grain within the marble conjures up images of mountain in full flight.

Huanghuali frame and stand with an absence of decoration pattern, the apron slightly wider than normal stand, end of oar shaped spandrels on the sides of the posts slightly tapering inwards, the feet of table screen in rectangular shape combined with opening of the base has a sturdy look.

The apron and stand are unembellished with clean lines and functional design.

024

黄花梨
镶绿石插屏

明末清初

高 50 厘米　宽 52.4 厘米　厚 20.8 厘米

绿石屏心，纹路不甚突出，一面较润而深沉，另一面较干涩而清爽，肌理丰富，但不具明显物象，略成山水图画，与常见纹路特征清晰明确的石板不同，别有一番意味。

素边框，不设绦环板，下为刀牙板，造型拙朴。立柱亦素。壶瓶式站牙。屏墩长方形，下挖亮脚形成矮足，即拱桥式。

有一批这样造型简洁、尺寸近方的黄花梨座屏或插屏，造型、风格趋同，制作于明清之际，此屏属于其中偏大者。

南宋
李嵩
《赤壁赋图》页 局部
美国纳尔逊-阿特金斯艺术博物馆藏

Southern Song Dynasty
Li Song
Part of Album of *The Red Cliff Ode*
The Nelso-Atkins Museum of Art

Huanghuali and green stone table screen

LATE MING TO EARLY QING DYNASTY
H 50CM W 52.4CM D 20.8CM

Huanghuali frame and stand without decoration pattern, the rectangular feet with opening of base has a sturdy look. There is a group of square shaped *huanghuali* frame table screens produced during the late Ming and early Qing Dynasty, this piece could be related with the group.

Unlike *dali* marble, the grains of green stone is less pronounced, portraying shapes which have a blended appearance creating a swirling water effect.

025

黄花梨
镶大理石插屏

清早期
高 71 厘米　宽 76.5 厘米　厚 24.5 厘米

中大型屏，非大案不足以陈。

屏心大理石，地子白中透黄，纹路黄绿、灰绿相间，深浅变化，图案抽象，有当代抽象水墨意趣，若雨过春山，葱茏满眼；又若春潮初泛，雨丝连连。两面可观，虚实各不相同。大理石中纹路泛绿者有之，但满眼皆绿且纹路深浅变化丰富者少见，夏日观之生凉，涤荡心胸，似乎都能想见制作者见此佳石后制成器具后的喜悦。

框、座以顺纹黄花梨制成，不雕饰或起线，甚平素。宽边框。屏座仅设横枨承接屏框。刀牙板，牙条中段垂下为洼膛肚式，较为别致。立柱平素，桨腿式站牙，拱桥式屏墩。

另见一件相同造型、石质的插屏，尺寸稍小，应为同一批制品。

明
戴进
《风雨归舟图》轴局部
台北故宫博物院藏

Ming Dynasty
Dai Jin
Part of Hangingscroll of *Coming Home in the Wind and Rain*
The Palace Museum, Taipei

HUANGHUALI AND DALI MARBLE TABLE SCREEN

EARLY QING DYNASTY
H 71CM W 76.5CM D 24.5CM

This table screen is of middle to large size, usually being displayed on large tables. The screen is removable and reinserted through a 'tongue and groove' construction.

The frame and stand made of *huanghuali* with clean unembellished lines. The frame is slightly wider than the usual standard which has a sturdy appearance.

The *dali* marble in the central panel has white colour with ink-greenish abstract grains through. The main feature of this marble is a pronounced ink-greenish grain occupying the major lower portion, rarely seen in *dali* marble.

026

黄花梨夔纹框
镶大理石座屏

明晚期

高 43.8 厘米　宽 44 厘米　厚 14.5 厘米

　　中型座屏，屏心大理石横幅，纹路为江山平远式，地子黄白色，若雾气朦胧，隔岸山列如屏，起伏轻缓，远处隐约又有一重，意境冲淡。背面几不见纹路，乳白色质地，两峰若有若无，岁月加持，形成温润且细节丰富的皮壳，颇耐寻味。

　　黄花梨框、座，皮壳正反有别，正面干黄，清爽宜人，背面黑褐色，深沉凝重。屏框下落膛装绦环板，阴刻双夔纹。屏下设牙板，牙头处勾回，如小足般落于地上。桨腿式站牙，转折方硬。座墩如卷几，两足向内勾卷如鸟喙。

　　绦环板的夔纹样式，与清代常见的夔龙纹不同，值得注意。其形若鱼，扁首，独角，尖喙末端卷珠状，背上有鳍状物，古拙质朴，如同直接从青铜器上摹刻而来，尚带原始气息。这种纹饰主要流行于明晚期至清初，可见于当时的铜器、瓷器等装饰中，是晚明仿古之风使然。至清康熙时，这种纹饰逐渐制式化，乾隆时变化更加复杂，虽雕琢更加娴熟，磨制更加精细，然神采已失，远不及早期之朴拙生动。

　　一般所见插屏下的牙板皆是悬空，与地面有一定距离，此牙板的牙头触地，亦是较特殊做法，以至初见此物，怀疑屏墩是否因损伤而磨矮，然细查后知原即如此。揣测其意，或是即如此设计，如同小足着地，与屏墩三足鼎立，提供支持。或是其下原有活动的"工"字形地栿，牙头和屏墩落于其上，本书收录黄花梨花卉纹框镶大理石座屏（件47），其下带有可活动的"亚"字形地栿，披水牙板与屏墩下端平齐，与此件情况相仿。

　　另见有黄花梨夔纹框镶石刻《孝经》座屏，刻有明崇祯款识，高43.7厘米、宽35.5厘米、厚13厘米，无论整体造型还是绦环板装饰，都与本例十分相似，足可证此屏制作时间，亦可证此屏屏墩原即如此。

　　目前所见，有这样一类黄花梨制座屏，尺寸多在40厘米上下，多见方，个别尺寸更大或长方者。屏框平素，宽厚适度。有的设绦环板，纹饰朴实。下方多设素直牙条，牙头做成刀牙，少见者如本例。站牙为桨腿式，转折较方直。屏墩有的是平足落地，有的是内卷足落地。其榫卯结构亦基本相类，加工干脆利落，法度谨严。这类座屏，制作时间在明末清初，可能是同一地区同一流派工匠所为。

177

黃花梨夔龍紋框鑲大理石座屏

郭河陽趙松雪皆有平遠圖傳世遠山邃隱雲炯氳氲有虯龍藏焉松雪自謙曰余自幼學書之餘時戲弄小筆然於山水獨不能工蓋自唐以來如王右丞大小李將軍鄭廣文諸公奇絕之迹不能二見至五代荊關董范輩出皆與近世筆意遼絕畫之所由造化心源古今亦然細審是屏質凝如脂橫列雲山三痕又若重山無數炯靄迷離變化莫測近慮沖淡簡練中景厚重且不乏空靈遠山綿延不絕暗藏玄機此幅當與河陽松雪之意甚契
己亥年冬月為蕉園藏屏題劉丹

刘丹
跋"黄花梨夔纹框镶大理石座屏"
2019

Liu Dan
Postscript of '*Huanghuali* framed *dali* marble table screen'
2019

宋
李氏
《潇湘卧游图》卷局部
日本东京国立博物馆藏

Song Dynasty
Artist named Li
Part of Handscroll of *Landscape of Dongting Lake*
Tokyo National Museum

Huanghuali framed *dali* marble table screen

Late Ming Dynasty
H 43.8cm W 44cm D 14.5cm

This table screen is medium size, the white brownish *dali* marble has smoky ink grains crossed in the middle and appears as an archaic landscape painting with a mountain behind a river covered by mist. The reverse side marble is smooth cream without grains.

A pair of ancient mysterious animal *kui* are carved onto the panel below the frame. The *kui* had fish body and fin on its back, with flat head with horn. This depiction is close to the *kui* from bronze vessel or ceramic objects which was a popular decoration pattern during the late Ming Dynasty.

The toes of the apron touch the ground. In general, this is very unusual finish for the foot and toe. When we first inspected this unusual treatment, we suspected it could be caused by damage of the foot, however through careful examination, we found this is as originally intended. It is also possible that is was part of a removable H-shape trestle stand. There is a *huanghuali* and *dali* marble table screen has a removable stand discussed in this collection.

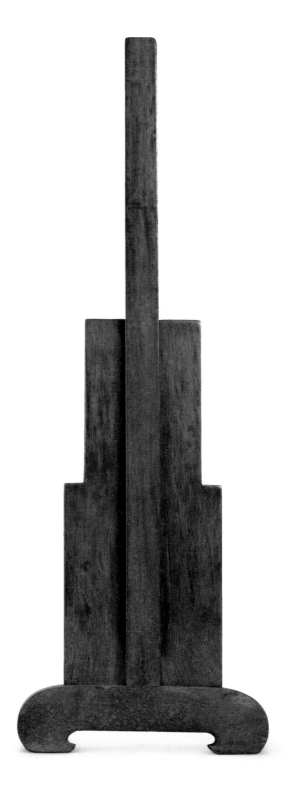

027

黄花梨
镶绿石座屏

清早中期
高 29.1 厘米　宽 24.5 厘米　厚 14.2 厘米

  中小型屏。屏心以深浅不同的绿色石纹构成图画，层层相叠，如春水荡漾，波光粼粼，中下方则磨制成两个对坐人物，颇有江水浩渺、随波浮游天地间之感。这种数层连弧状石纹是绿石屏心较为典型的一种形式，利用石头本身的丰富层次，采用波浪式磨制，形成上下起伏的纹路。背面磨平，不特意追求纹路，却在不经意间形成云海浩渺的图画，亦可观。

  黄花梨为框、座，红皮壳，形成浑然一体的效果。下设绦环板，上有笔管式开光。壶门式披水牙板，桨腿式站牙，造型甚简，只沿边铲阴线装饰。拱桥式屏墩，上方转角处小委角，阴刻卷珠纹点缀。

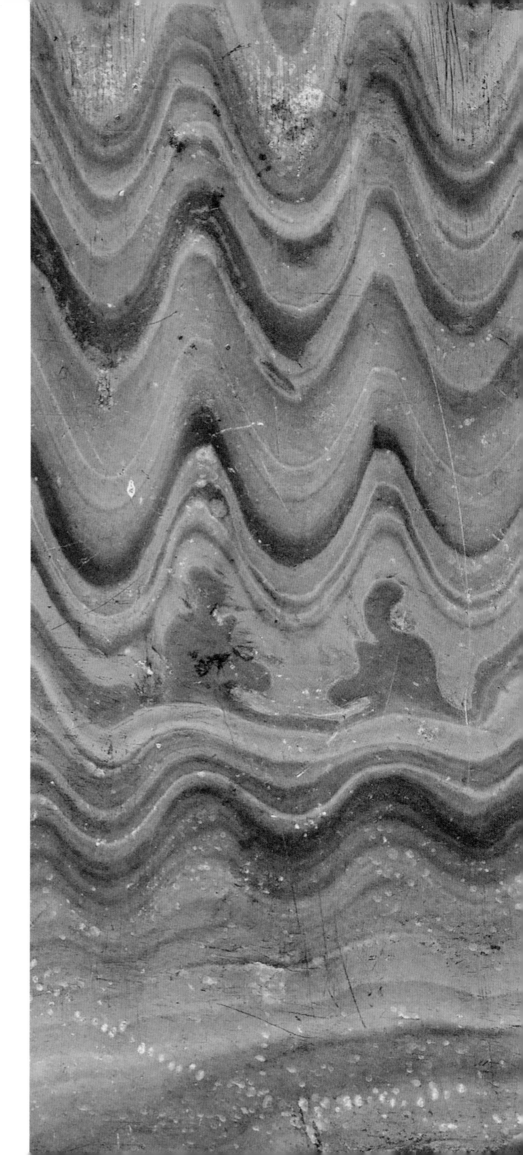

183

Huanghuali and green stone table screen

EARLY TO MIDDLE QING DYNASTY
H 29.1CM W24.5CM D 14.2CM

The table screen of mid to small size. *Huanghuali* frame and stand in dark brown, open carved panel on top of scrolling cusped apron joined on upright brackets on the sides, stand on curved shaped feet.

The central panel has various shades of green grains like water waves, two carved figures were seated face to face on top of water waves. These wave-grains are a typical feature of green stone which, through a grounding technique, makes heavy waves of the natural shades.

028

黄花梨
镶大理石座屏

清中期
高 22.5 厘米　宽 17.2 厘米　厚 9.6 厘米

　　此屏尺寸小巧，为砚屏之属。

　　屏心大理石仅巴掌大，然气象万千，纹路如云气氤氲，山色空蒙，极润泽，观之生酣畅淋漓之感。黑色石纹若墨之晕散，恰其又为砚旁之物，令人幻想这莫不是砚中之墨浸染而成？石涛有《云山图》（故宫博物院藏），山势磊落，云气变幻极佳，与此小屏自然图画甚契。徐霞客见大理石赞叹"故知造物之愈出愈奇，从此丹青一家，皆为俗笔，而画苑可废矣"，未免有所夸张，但此屏大理石纹路与山水画的意境互相阐发，却无不可。大理石两侧可观，山势一侧稍雄浑，另一侧空灵。石片中间微鼓，四周略低，更觉浑圆可爱。

　　黄花梨框、座，较厚重的北方黑红皮壳，润而亮。绦环板亦鼓起，上有笔管式开光。壶瓶式站牙，拱桥式墩。

　　此屏造型不属突出者，然尺寸小巧，石纹妍美，令人见之喜爱不已，难以释手。

HUANGHUALI AND *DALI* MARBLE TABLE SCREEN

MIDDLE QING DYNASTY
H 22.5CM　W 17.2CM　D 9.6CM

Table screens of this size are called *yanping*, which function as an inkstone screen and shield ground ink from wind. Refer to the 15th century handscroll by Xie Huan, *A Literary Gallery in the Apricot Garden*, in the collection of the Metropolitan Museum of Art, New York, where a similar screen is placed next to an inkstone. Surviving examples of inkstone screens made in *huanghuali* are very rare.

清
石涛
《云山图》轴
故宫博物院藏

Qing Dynasty
Shi Tao
Hangingscroll of *Clouds and Mountains*
The Palace Museum

029

紫檀
镶大理石插屏

明末清初
高 57.5 厘米　宽 53 厘米　厚 23 厘米

　　此屏属中型偏大者。屏心为大理石，皮壳老辣，地子已为黄白色，纹路黄绿相间，为平远山势，如浓雾环山、春河浮冰，意境空灵，属大理石屏中出类拔萃者。

　　屏框、座分体式。屏座的立柱与屏框交接处设在站牙的边缘处，外观平齐，融为一体。屏框宽阔，紫檀质感肃穆，衬托屏心。绦环板上为通长的笔管式开光，通透醒目，使人在欣赏屏心时也不至忽略精彩的披水牙板。全器他处皆素，唯披水牙板点缀倒垂如意云纹，采用浅浮雕手法，转折圆润，起伏柔和。站牙为桨腿式变体，中有长圆形开光，边缘皆圆润。拱桥式屏墩。

　　明式的紫檀插屏比较少见，整体造型颇有明味。

紫檀鑲雲石插屏

有清一代山水諸家承緒唐宋師師灑元明然得其三昧者寧～於眼格能得真絰者亦鮮董思翁雲畫家以古人為師已自上乘進此當以天地為師每朝起看雲烟變幻絕近畫中山傳神者必以形形與心手相湊而相忘神之所托也畫當如是造化自然雲石為靈氣所聚結而為象示現之形豈非神所托乎此屏石質細密而不失天氣選色絲繁却得不俗屏中遠山脈脈霧海隱隱如瑤池倒瀉沉浮自如克紹松江高格自成殆非痴人所能想見

己亥年冬月為蔗園藏屏題 金陵劉丹

刘丹
跋"紫檀镶大理石插屏"
2019

Liu Dan
Postscript of 'Red Sandalwood and *dali* marble table screen'
2019

Red sandalwood and *dali* marble table screen

Late Ming to Early Qing Dynasty
H 57.5CM W 53CM D 23CM

This table screen of middle size, light brown-white *dali* marble form the central panel with green and yellowish grains, resembling an ancient landscape painting. The *dali* marble mounted in red sandalwood stand as table screen. The framed panel is removable from the stand. The beautiful colour between red sandalwood and *dali* marble in an elegant feature. An oval shaped opening inserted in the bottom rectangular panel decorated simply with carved cloud pattern on the apron.

195

030

黄花梨
镶大理石插屏

明末清初

高 63 厘米　宽 46 厘米　厚 27.5 厘米

　　屏心大理石，山脉一重，若画笔随手挥洒而成，白黄色地子，上有深浅条纹，若云气弥散。

　　屏框、座黄花梨制，正面干黄色皮壳，甚为洁净、清爽。屏背黑皮壳，沧桑老辣，甚为古旧。边框较宽，外沿铲地起皮条线，一直延伸至立柱，与下枨交圈，形成大气、自然的装饰效果。屏座上设双绦环板，内为剑环式开光。壸门式披水牙板，牙头处镂为卷云纹，曲线温和。立柱内有槽口容纳屏框上的榫舌，外形则与屏框自然交圈。桨腿式站牙，亦沿边起宽皮条线与屏框呼应。屏墩拱桥式，镂挖亮脚甚高，形成两个内外皆撇小足的高足。

　　此屏屏心近方，但下方绦环板、屏墩皆高，整体为竖屏式，形成稳重挺拔的效果。造型落落大方，比例适度，线条简洁明快，观之如堂堂君子，搁置案头，可正人心。

HUANGHUALI AND DALI MARBLE TABLE SCREEN

LATE MING TO EARLY QING DYNASTY
H 63CM　W 46CM　D 27.5CM

The screen in vertical rectangular shape. White yellowish *dali* marble panel has various shades of grain gives the appearance of moving clouds. The frame of the panel and stand are made of *huanghuali*. Wide flat-beaded border moulding on the frame continuous flow into the stand. A pair of open carved panels below. The ornamental apron opening has carved cloud decoration. The base footing has a wide stance with the toes on each foot flared outward in opposing directions affording excellent stability.

031

黑漆撒螺钿框
镶大理石座屏

明末清初

高 50.5 厘米　宽 39.5 厘米　厚 20 厘米

　　座屏状态极佳，原始皮壳，整体一气呵成，浑然一体。

　　屏心大理石，正背可观，正面灰白皮壳，其上有山纹几重，是以人工磨制而成，制作手法高超，"虽由人作，宛自天开"，与山水画意境相通。背面自然成图，青灰色石纹，周匝山峰罗列，中间若云气。

　　框、座棬以致密黑漆，发色纯正，质地坚硬，撒螺钿装饰，因年久，部分螺钿片掉落，空留印痕，如雪花落地，反增肌理变化。边框上方做成大圆角，柔和而内蕴力量。与众不同的是其上方边框远宽于其他部位，与下方绦环板、牙板形成对应关系，反觉更加协调、稳重。

　　屏座设绦环板二，海棠形开光内镶嵌大理石板，与上方屏心呼应。屏下设壶门式披水牙板，两侧波折后有翅状勾卷云纹，这种样式的牙板，出现于明清之际，盛行于清早期，以康熙时期最为典型。屏墩和边框间设连弧状斜枨，为简化变体的夔龙纹。屏墩和边框间的支撑构件，自宋时出现斜枨和站牙两种样式，此属前者。屏墩形如几座，转折柔和，两足外撇上翘，内侧挖成如意形亮脚。站牙、绦环板开光、披水牙板、屏墩边缘皆起碗口线装饰。

　　此屏整体造型简洁明快，设计得体，曲线运用娴熟，可见匠人高超的技艺和独到的审美品位。出自扬州地区，为传世座屏之经典。

Black lacquer mother-of-pearl framed *dali* marble table screen

Late Ming to Early Qing Dynasty
H 50.5CM   W 39.5CM   D 20CM

This piece may be from the Yangzhou area. It is in overall good condition with mountain images carved into the white-grey marble on the obverse side, natural grey grains forming the reverse side.

Minute pieces of mother-of-pearl have been lost from the black lacquer frame and stand through the years leaving marks on the lacquer surface. These small marks give the surface a texture of antique elegant beauty.

The upper panel of the frame is wider than the side and bottom panels of the frame. The top two inner corners are finished in a curved angle giving the visual effect of expansive outward radiation whilst the bottom two corners, at right-angles, form a base from which the mountains rise. The *dali* marble panel surround by quatre-lobed oval opening on the two carved smaller panels set inside a rectangular carved panel. The feet of the stand are embellished with outwards flaring toes, tips turning upward. The apron has cloud scrolling bounded by a raised perimeter line at the margin. This type of apron first appeared in the late Ming Dynasty, becoming popular in the early Qing commonly identified with the Kangxi reign.

南宋
直翁（传）
《药山李翱问道图》卷局部
美国纽约大都会艺术博物馆藏

Southern Song Dynasty
Attributed to Zhi Weng
Part of Handscroll of *Li Ao's Search for Taoist Truth*
The Metropolitan Museum of Art, New York

032

楠木
镶大理石座屏

清早中期
高 67 厘米　宽 68 厘米　厚 34 厘米

屏心大理石，黄地，深褐色石纹如流云聚散，只余下方一带白如玉，其纹于中部略凝结为山，左侧耸立一峰，若湖石冠云，画面清新跳脱。屏背深色，不显纹路。

屏框、座楠木制成，原皮壳，观感、触感皆温婉。延续屏心富于装饰的效果，两个绦环板上镂雕朵花纹，三花相连，中为梅花，旁为菱花，剪影效果明显，宛若窗花。下设披水牙板，壶门式，中有分心花，沿边阳线至两端衍为卷叶状。屏框为双面工，背后皮壳老辣。壶瓶式站牙，内侧镂为卷叶纹，下承抱鼓墩，鼓面雕渠花瓣。

屏心纹路奇诡，颇俏丽，玲珑的绦环板装饰为之增色不少。

205

NANMU AND DALI MARBLE TABLE SCREEN

EARLY TO MIDDLE QING DYNASTY
H 67CM W 68CM D 34CM

The rich golden *nanmu* frame enclosing the attractively variegated black, white and grey *dali* marble panel. Free relief flower pattern on carved panel, ornamental opening apron, raised on a circular drum-shaped trestle base.

Two lattice-work decorations with two plum and one water chestnut flowers form the bottom of the screen. The apron of the base is finished in a scroll effect with raise border forming a line joining in the middle, upward pointing where each side of the raised boarder meet.

Violin shaped spandrel attached to a base displaying two drum shaped carvings support on a simple trestle.

033

黄花梨双螭纹框镶绿石插屏

清中期
高 44.5 厘米　宽 40 厘米　厚 18.5 厘米

屏心用石比一般绿石更细腻，其上有绿、黄、紫三色纹路。一面如千山叠翠，紫云浮空；另一面如飞霞经天，彩光万道。纹路变化之奇特，物象之抽象，色彩之奇瑰，皆是他石所少有。

黄花梨为框，有透雕双螭龙纹绦环板和披水牙板，雕刻手法娴熟，线脚利落，雕刻爽利，披水牙板上阴刻边线、卷草纹，简括流畅，数刀即成。拱桥式屏墩，作变体双首共身夔龙形，只雕出简练的双眼。

从螭龙纹形象及加工技术的成熟度来看，应为清中期前后制品。其雕刻风格与福建地区一些黄花梨围屏上所见接近，不排除为闽作家具。

HUANGHUALI FRAMED GREEN STONE TABLE SCREEN

MIDDLE QING DYNASTY
H 44.5CM　W 40CM　D 18.5CM

The stone panel has unique characteristics of a rich colour and smooth texture, it is close to the green stone colour. A pair of free relief carved dragons on the opening panel. A scrolling vine like pattern carved on a shapely bottom apron.

The details of the carved dragons may date this piece from middle to late Qing Dynasty. The carving technique indicates this piece may be from the Fujian area.

034

黄花梨螭龙纹框镶绿石山水人物图座屏

清早中期

高 54.5 厘米　宽 44 厘米　厚 26 厘米

　　屏心为绿石，极润泽，石纹根据不同深浅的绿色层次磨制而成，两面可观。正面为山水人物图，一日当空，旁有云翳，山脉两重，自画面左上角斜伸向右下角，作三山耸立状，又有人物点缀其间，或躬手而立，或乘舟浮游，或腾云俯观，周匝若水泄云奔，中含物象，影影绰绰，奇诡难测，清奇雅丽，如临仙界。背面较浑浊，中部磨显一兽，若虎若猫，为绿石常见之装饰手法。

　　黄花梨框、座，干黄色原皮壳，甚为洁净。屏框看面打洼，转角做委角线，背面另起子框，设暗销，屏心可活拆。绦环板镂空雕双螭龙纹，拱卫一枝灵芝。壶门式披水牙板，铲地浮雕缠枝宝相花纹，雕刻下刀婉转波折，铲地亦深，立体感甚强。站牙以灵芝纹雕成，布局巧妙，神采飞扬。抱鼓式屏墩，中间浮雕变体兽面纹，鼻头衍为倒垂如意纹，下挖壶门式亮脚。抱鼓的鼓面铲地浮雕杂宝纹，计有葫芦、海螺、扇、卷轴、如意、拍板、犀角、书剑，各系飘带。

　　从缠枝宝相花等纹饰特征看，此屏当为清中早期之物。地域特征不甚明显，闽作、苏作皆有可能。屏心甚佳，堪为绿石屏之代表。装饰纹路繁而不杂，雕工甚精，尤其是灵芝纹站牙设计予人意外的惊喜，干爽皮壳亦为之增色不少。

Huanghuali framed green stone table screen

Early to Middle Qing Dynasty
H 54.5cm W 44cm D 26cm

The ink green coloured green stone has very fine smooth texture. The obverse side depicts mountains and water with human figures whilst the reverse side imagines a mysterious animal.

Huanghuali frame and stand in brownish colour. The back of frame has a hidden lock to allow removable of the central panel. Three-dimensional relief and open carved dragon on the carved panel. Excellent carving technique is displayed in the shaped apron into which is carved a raised scrolling flower pattern. A three-dimensional *lingzhi* pattern is carved into the spandrels. Circular drum decoration with auspicious patterns on the base stand.

The method used for the scrolling flower pattern indicates this piece may be made in the early or middle Qing Dynasty from either the Jiangsu or Fujian region.

035

紫檀嵌玉棂格
镶大理石座屏

清早中期

高 27.3 厘米　宽 24 厘米　厚 12.5 厘米

横幅大理石屏心，乳白色地，远山一重，黄绿色纹，若春山新绿，洁净青翠，近处又有淡色石纹，如山似水。背面纹路如淡墨轻扫，点画若隐若现。一般所见，明代大理石，纹多黑如墨，或有灰黑色，以古朴见长。清中晚期挂屏所用大理石，纹多灰黑相间，以奇诡见长。黄绿色纹大理石少有，其纹清新淡雅，别具一格。

紫檀框、座，造型奇巧。屏框攒框后下方又设一体的小框，形成一个倒"凸"字形，小框内设一排计10个八棱形白玉棂格。两旁"凸"字形站牙，上方挖槽，夹嵌屏框，下方开小亮脚，与上方呼应。屏框的下帐另出榫舌，插入阔厚的牙板内，紧密安稳。长方形屏墩，拱桥式。

此屏白玉、大理石、紫檀三种材质，皆为细腻、珍贵之物，三者搭配适宜，各擅所长。其造型端庄，不设一纹，加工谨细，兼得富贵、文雅之趣。清宫旧藏有清早中期紫檀嵌玉棂格罗汉床，亦是以玉棂格装饰围子，与此屏意趣相合。

紫檀鑲玉柱大理石座屏

明清雲石之賞當以阮文達公為首座每得一佳屏俱為命名且賦詩銘之以示珍愛後世之好事者皆以此為模範雲石中之可觀者當以品格定高下意與古會乃為上品此屏底色如凝脂約略幾抹黛山神似趙千里希遠伯仲筆殆非俗物内府藏萬松金闕一卷足可證與雲林子有詩以副其後玄萬松金闕郁岌巍望望人間思沉寥留得前朝金碧畫傖人天際若為招趙文敏亦言希遠所作清潤雅麗自成一家亦近世之奇也買之當下亦然
己亥年冬月為蕉園藏屏題金陵劉丹并識

刘丹
跋"紫檀嵌玉根格镶大理石座屏"
2019

Liu Dan
Postscript of 'Red sandalwood, jade and *dali* marble table screen'
2019

Red sandalwood, jade and *dali* marble table screen

Early to Middle Qing Dynasty
H 27.3CM W 24CM D 12.5CM

Cream white *dali* marble has green and yellowish grains resembling beautiful mountains in early spring. On the reverse side of the marble, the natural grains look extremely fine, diffuse dots creating a blurred, atmospheric effects. In general, the marble with grey and black colour grain was often used on central panels during the middle to later Qing Dynasty, however the green and yellowish grain are rare colours.

Red sandalwood frame and stand presented special ideas for design. There is a smaller frame under the main frame, within which there are ten pieces of jade columns appearing to support the marble above. The larger frame fitting into clefts in the base in a cradle like manner.

Employing three precious materials on one object is relatively rare. Another piece using different precious materials is a couch bed from the Qing Imperial Collection in the Palace Museum.

036

红擦漆卍字纹座
大理石插屏

清早期
高 52 厘米　宽 73.6 厘米　厚 18.3 厘米

　　大理石石板一块，附座而成。大理石深浅不同的石纹布满画面，如同绞胎效果，形若舆图，岛屿、湖泊星罗棋布，极富变化，一面润泽，另一面干涩，意趣各不相同。

　　屏板两侧下方将前方剔薄一条为榫舌，纳入屏座，属较细致做法，一般石板多直接插入屏座，不单独做榫舌。

　　屏座榆木薄擦红漆，漆质较粗，木纹隐现。设绦环板三块，委角长方形开光内镂雕打洼卍字纹，中间较疏，两侧较密。下设壸门牙板，沿边铲利索的阴线。立柱头圆雕仰莲纹，下托以荷叶，雕饰简练大方，立柱断面作委角方形。站牙为螭龙纹盘绕而成，造型天真随意。屏墩形如矮几，两侧内收后微撇如蹄足，下方挖成壸门式亮脚，边铲阴线。

Cinnabar lacquer and *dali* marble table screen

Early Qing Dynasty
H 52cm W 73.6cm D 18.3cm

The *dali* marble presented in rich grains, like an ancient map, the water coursing through different islands. The lower sides of the marble panels have mortises sunk into each side making the panel fit into the stand. The three latticework sections contain symbols of auspicious design placed within two supporting crossmembers.

The stand made of elm coated cinnabar lacquer, the wooden grain being visible in the lacquer coating. Three carved panels with auspicious patterns are below followed by a shaped apron. There are lotus flower decorations at the head of the posts. Dragons decorate the spandrels with the base feet splaying outwards for support.

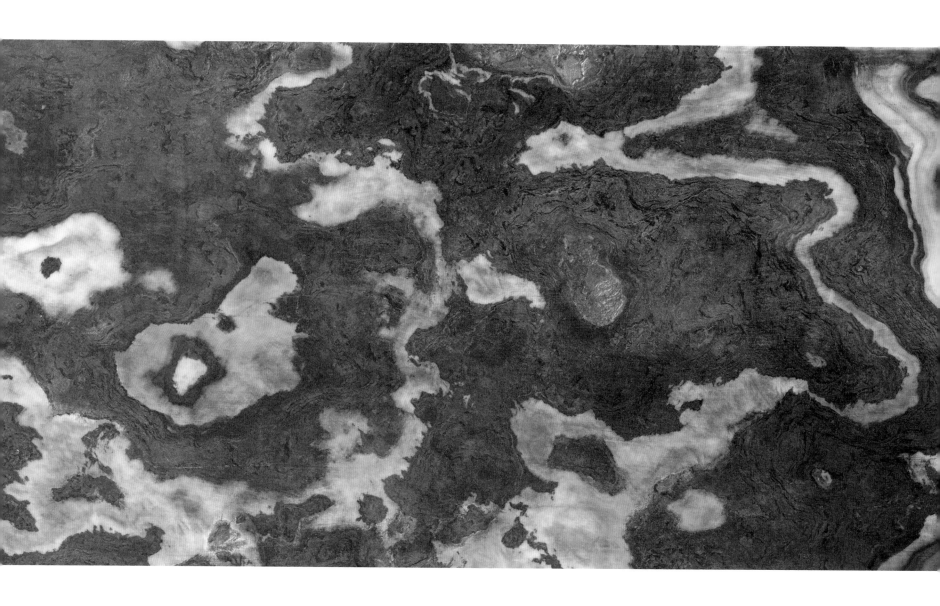

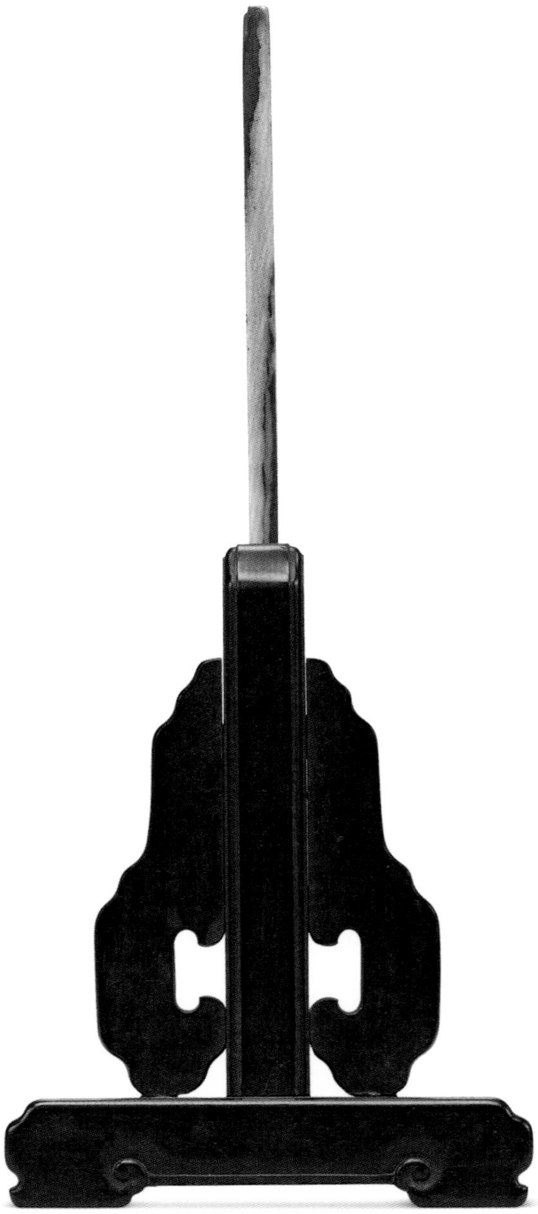

037

乌木座
大理石插屏

清中期
高 36.5 厘米　宽 26 厘米　厚 15 厘米

　　大理石厚板为屏，不设边框，附座而成，是名副其实的"插牌子"。

　　两面可观。一面白色地子上黑、绿色纹路，集中在左下角，余处洁净无纹。纹若春潮初泛、万马奔腾，观之令人酣畅淋漓。黑色深沉，绿色清新，白色如玉，三色分明。另一面石纹迥异，黄白色地子极润泽，布局空灵，只上方周缘有黑色石斑，下方有绿褐色石纹若山峰，有大江东去、天地浩渺之感。

　　大理石石纹千变万化，各具特色，以妍美而言，此例为其中翘楚。

　　屏座乌木制，取其乌黑细腻，可衬托白地黑章的大理石。设绦环板、披水牙板，皆素朴。站牙造型稍有特点，若简化夔纹的轮廓，局部有柔软的连弧状造型，勾卷却颇锐利，这种造型意趣，使人联想到本书收录的另一件乌木镶大理石座屏的屏墩。

Wumu and *dali* marble table screen

MIDDLE QING DYNASTY
H 36.5CM W 26CM D 15CM

A frameless cream white rectangular marble plate in vertical position is joined into a *wumu* stand. One side of the marble contains black and green grains principally located on the low left section. The reverse side of the marble panel features a beautiful colour combination of contrasting cream white, black and green clouds.

The *wumu* stand is of very fine texture, particularly the shining wooden combined with the cream white *dali* marble present interesting colours which related to another piece collected in this book.

038

黄花梨
镶大理石插屏

清早中期
高 32 厘米　宽 31.5 厘米　厚 16.5 厘米

　　横屏，中小型。屏心镶嵌大理石而成，单面为一斜向深色山脉自左下贯穿至右上，其形若带，蜿蜒崎岖，余处空中作乳黄色，层次丰富，本有横向的数道细碎裂纹，年久形成淡灰色石纹，宛若云浪，又若雪山重重。下方为灰青色山脉。

　　扁阔屏框，上方转角做成大圆角，外侧平直，内侧做成混面与屏心相接，这种做法多是清代紫檀、红木家具上的手法。屏座设壸门牙板，沿边起线，正面的边线衍为卷草纹。立柱造型简素，顶端转角处抹圆。变体螭龙形站牙，仅大略镂出轮廓。屏墩若小几，转折处有委角，外撇小尖脚，轻巧落于地上，屏墩内侧锼挖亮脚。

Huanghuali and *dali* marble table screen

EARLY TO MIDDLE QING DYNASTY
H 32CM W 31.5CM D 16.5CM

The screen is in a horizontal rectangular shape, with the *dali* marble as the panel. There are several hairline cracks across the middle of the marble, through time, the brownish colour has bled into the marble creating an appearance of mountain waves.

The top corners of the moulding used to hold the marble screen are rounded. This is a common technique used on red sandalwood or *hongmu* furniture during the Qing Dynasty. A scrolling grass pattern with raised edges forms the apron borders. A dragon shape forms the spandrels on top of a splayed base.

039

紫檀
镶大理石插屏

清中晚期
高 32 厘米　宽 29.5 厘米　厚 13.3 厘米

屏心洁白，有淡淡山脉一重，正背可观，意境悠远。

这种白色地子浅淡纹路的大理石屏，在清中晚期时应用较多。故宫博物院藏有一件楠木镶大理石面条桌，大理石纹亦甚清淡，上有题诗一首："倚玉尤含润，看山恰有光。何如盈丈地，云气浩茫茫。"正可见古人对这类石纹的鉴赏态度。

紫檀框、座。屏框内做大圆角，看面中间隐起皮条线，制作谨细。屏座镶落膛起鼓绦环板，披水牙板光素，边缘挖壶门线。直立柱，变体壶瓶式站牙，立柱和站牙侧面皆打洼装饰。拱桥式屏墩，上方转角处挖大委角。

此屏深色框、座与白色屏心相映衬，一抹云山为其增添意趣。

235

清
梅清
《黄山十九景图》册之二
上海博物馆藏

Qing Dynasty
Mei Qing
One of Album of *Nineteen Sights of Huangshan Mountain*
The Shanghai Museum

Red sandalwood and *dali* marble table screen

Middle to Late Qing Dynasty
H 32cm W 29.5cm D 13.3cm

White *dali* marble has mountain shaped grains located in the middle to lower sections of the panel. This type of white marble with light touch grains was popular during the middle to late Qing Dynasty. A similar piece is a *nanmu* and *dali* marble table in the collection of the Palace Museum.

The red sandalwood frame, with beautiful colour contract between the red sandalwood and white marble presents a fine elegant look enhanced by the rounded inner corners of the frame.

040

黄花梨螭龙纹框镶绿石插屏

清早中期
高 42.5 厘米　宽 39 厘米　厚 19 厘米

屏心绿石，黄绿色纹路相间，两面可观。一面图案如海中浮有仙山，山峦丛列，层层不穷，中有三山，山顶皆有人物形象，人大于山，轮廓简练、写意，作躬身作揖状，想为海中神仙，近处波浪中礁石突兀，其周有物若海兽，或攀爬，或沉浮，光怪陆离。另一面图案较为抽象，若川流云涌。

屏框素直，宽窄适中。屏座绦环板上圆角长方形开光，内浮雕螭龙纹，样式奇特，两旁螭龙拱卫着太极图，螭龙为双龙共体式，向内的螭首较大，长喙独角，杏核眼，前足为卷叶状，但回卷的叶片上又雕圆珠，形成简略的凤首纹，且与太极图呼应。向外的螭首较小，生于尾处，造型亦稍简。壶门式牙板，正中浮雕相背的象首瑞兽，阳线至两端牙头处衍为回纹。变体螭龙式站牙，尾翼回卷抵住立柱，其身躯上又阴刻螭首，尾部亦然，为双螭相吞状，整体其实是由三螭纹组合而成。立柱简洁无纹。屏墩两头阴刻为螭首相吞状，下有挖出的矮小撇足。

屏座上的图案雕饰较粗率，铲地不平整，甚至仅沿阴刻线条稍斜铲即可，与细腻精致的江南地区雕刻不同，或是山西等地北方匠人所制。

Huanghuali framed green stone screen

Early to Middle Qing Dynasty
H 42.5cm W 39cm D 19cm

The green stone has green and yellowish grains on both sides, one of which has images of mysterious mountains, with the sages or gods atop the mountains and with the sea and rocks located at the bottom of the screen. The reverse side panel features floating clouds.

The frame is simple and undecorated. Below is featured a pair of dragons in relief carving on the rectangular panel. One dragon being slightly larger with an elongated mouth, a single eye and a horn. In the centre between the two dragons is carved a pearl supported by scrolling leaves indicating it may relate to *Taijitu* . Auspicious animals are carved on the rectangular shaped apron with spiral decorations at the side. The spandrels are carved in the shape of standing dragons, their tails positioned against the post. As the carving technique on this piece is not as fine as objects produced from Jiangnan region, it may be from Northern China, such as Shanxi province.

041

紫檀
镶大理石插屏

清中期
高 20.5 厘米　宽 21.8 厘米　厚 10.5 厘米

　　小型插屏，造型工整。

　　屏心大理石，白地上灰色石纹，形成山石磊落、云气弥漫的近景山水，与常见烟雨缥缈的米家山水不同，更似王履《华山图》册中所见，具写实山水面目，别有一番趣味。正背可观，有趣的是一面屏心上原有红漆字竖行，可辨为"祖国河山大……人民力量强"，为近几十年所题，颇具时代特色。座屏作为座右之器，题写铭文于其上是重要的用途，老屏新铭，此屏可充一例。

　　屏框、座紫檀制成，屏框简素，屏下绦环板上设有笔管式开光，下方牙板甚厚，曲尺状轮廓，沿边起阳线。立柱头雕为方尊式。站牙轮廓方折，为变体如意式。屏墩直方，下方挖低矮狭长的亮脚。

明
唐寅
《为款鹤先生写意图》卷局部
上海博物馆藏

Ming Dynasty
Tang Yin
Part of Handscroll of *Freehand brushwork painting for Mr. Kuanhe*
The Shanghai Museum

Red sandalwood and *dali* marble table screen

Middle Qing Dynasty
H 20.5CM W 21.8CM D 10.5CM

White marble with grey grains similar to landscape painting. On one side of the marble can be seen traces of a slogan written in red paint 'Mountain and River in Great Motherland, Great Power People'. Calligraphy is a frequently used method for decorating table screens. People from differing periods had different views on ways to decorate table screens, this is an interesting example.

The frame and stand are made from red sandalwood. The panel below the screen is carved with a narrow oblong opening. The bottom edge of the apron has raised boarder. The spandrels are heavy shaped in hard angles with the footings in similar hard shape.

042

黄花梨
镶大理石插屏

清早中期
高 32.8 厘米　宽 27.5 厘米　厚 16.5 厘米

　　屏心大理石，纹路天地分明，上方白色地微泛黄绿，皮壳丰富；下方淡灰色石纹，无他色，甚素气。石纹作山峦起伏状，如细笔勾勒，轮廓较肯定，造型奇诡，若墙壁水渍般变化无端，质感轻柔，宛若笼纱。山体若以卷云皴、披麻皴法描出，再略以淡墨渲染状，颇合笔墨情趣。屏心两面相似，又各有其味。这种清雅的灰黑色纹路大理石，甚为少见。

　　屏框、座黄花梨制，为北方常见的枣红色原皮壳。绦环板上开鱼门洞，轮廓变化丰富，与石纹似有呼应之意。变体夔龙式站牙。抱鼓式屏墩，下开壸门亮脚，两足略呈花叶状，末端微翘，活泼自然。

Huanghuali and *Dali* Marble Table Screen

EARLY TO MIDDLE QING DYNASTY
H 32.8CM W 27.5CM D 16.5CM

White *dali* marble with slight green and yellowish grains at the top with light grey grains on the lower section. These grains form images of mountain groups of particularly rare colours for *dali* marble.

The frame and stand are made of *huanghuali* in a dark red colour. The stand is especially ornate and detailed comprising many flowing curved surfaces and shapes. The panel below the screen has narrow pierced opening with dragon shaped spandrels and circular drum-shapes.

043

黑红漆
镶大理石座屏

明中晚期

高 57 厘米　宽 53 厘米　厚 26.5 厘米

　　中型座屏，造型简练大方，髹黑、红两色漆，饶有汉风。

　　屏心大理石近方，两面可观，地子微黄，青黑色石纹若皴擦而成。正面为巨嶂般耸立的山峰，前后略作两层，纹路若墨笔渲染，浓淡变化。背面为更奇诡的山峰兀立，远处重重叠叠，若无尽头。正、背纹路有聚、散之变化。

　　屏框宽素，髹黑漆，密布细断如鱼鳞。其内子框，稍落膛，髹红漆。抱鼓墩式，为云气托鼓状，下设圭角。对石榴头站牙，下方翻简练的勾云纹。披水牙板壶门式，曲线波折，两端翻半如意云头。

　　此屏比例匀停，装饰得体，壶门曲线和如意云头纹造型，具明中晚期特征。

Black and cinnabar lacquer *Dali* marble table screen

Middle to Late Ming Dynasty
H 57CM W 53CM D 26.5CM

The central part of the screen is made of near square marble with yellowish and black veins. The front depicts towering peaks while the back has more peaks standing in a seemingly endless chain. The frame of the screen is unembellished with black lacquer and fine patterns while the inner frame is in cinnabar lacquer. The stand has a drum and cloud shaped design as the main body support on each side rested on a platform.

The screen has a balanced proportion and appropriate decorations, the curved line of apron and clouds shape has characteristics attributable to the Ming Dynasty. Apron with curved lines and a cloud head design are also characteristic of the middle and late Ming Dynasty.

044

红漆框
镶绿石座屏

明晚期
高 47 厘米　宽 48.5 厘米　厚 26 厘米

　　座屏中型尺寸，规矩稳重，是明清时期复框式座屏中之经典。

　　横幅屏心，黄绿色，正面若玄冰浮水，又如远山一重，云雾充塞。背面纹路较深，一片鸿蒙之态。

　　复框式，看面打洼带委角，复框间装剑环式开光绦环板，横向以矮老界为三段，竖向两开光并列，不设矮老。屏下壶门式披水牙板，牙头处为相抵卷云纹造型。卷云纹站牙，两边相合为变体石榴头状。抱鼓墩式屏墩，抱鼓下部出头并雕成卷云纹，即建筑斗拱上常见之麻叶云。其下为一体雕成的扁矮底座，如同须弥座上的圭角，上起压边线，下方为剔地壶门式，两头为相抵的卷云纹，典型明晚期特征。

　　屏的框、座皆髹红漆，布有小蛇腹断，发色极佳，温润雅正。左上角绦环板阳线处，露出银色地子，依此看他处磨损露出黑色地子者当是银色氧化而成。这种红漆以银色漆层为衬的做法，在一些佛像髹漆上也偶能见到，可能是发色所需，具体工艺和材料，还需请教方家。

　　此屏漆片采样经碳十四检测制作时间为1606年±21年，与起初断代相符，堪为研究明万历时期家具风格的典型实例。明万历时期的家具，造型更加规整，法度更加谨严，追求对称、均衡、稳定，开明式家具之先河，此屏正是这种风格的体现。

明
杜堇
《玩古图》轴局部
台北故宫博物院藏

Ming Dynasty
Du Jin
Part of Hangingscroll of *Appreciate Ancient Artifacts*
The Palace Museum, Taipei

Cinnabar lacquer framed green stone table screen

Late Ming Dynasty
H 47CM W 48.5CM D 26CM

Ornate wide perimeter frame containing eight carved panels, each decorated with elongated oblong shaped relief carvings. Apron with carved scrolling clouds with shaped end attached below the frame and carved scrolling cloud spandrels. The table screen of medium size was a typical late Ming feature.

The green and yellowish stone has water wave on the obverse side, the mysterious grains on the reverse side.

Beautifully finished cinnabar lacquer used on the frame and stand have fine cracks. Parts of the cinnabar lacquer has detached from the top left corner of the carved panel. The sliver based under the cinnabar has oxidised into black. The technique of using sliver based under cinnabar lacquer was also used on religious figures. This type of lacquer technique requires future research.

Carbon-14 dating 1606 ± 21 CE, This piece is typical of furniture from Wanli period. The Wanli period saw the introduction of a preference of more balance and symmetry in furniture which carried on into following periods.

045

黑红漆框
镶大理石座屏

明晚期

高 45 厘米   宽 45.5 厘米   厚 24.5 厘米

屏心大理石，两重山，若江南春雨初收，极润泽。

复框式，看面起剑脊棱式线脚，棱处为宽皮条线。框格内镶剑环式开光绦环板。壸门牙板，两头为对抵的卷云状。

卷草纹站牙，合则为变体石榴头式。抱鼓式墩，下设圭角。

此屏为复框镶绦环板式座屏之典型，盛行于明中晚期，与前件明晚期红漆框镶绿石座屏甚为相近。从其本身造型和装饰风格看，应出自山西中南部地区。其主体框架擦黑漆，装饰构件擦红漆，清新明快。屏心大理石属同类中之佼佼者。

五代
(传) 董源
《平林霁色图》卷局部
美国波士顿美术博物馆藏

Five Dynasties
Attributed to Dong Yuan
Part of Handscroll of *Landscape*
The Museum of Fine Arts, Boston

BLACK AND CINNABAR LACQUER *DALI* MARBLE TABLE SCREEN

LATE MING DYNASTY
H 45CM W 45.5CM D 24.5CM

Dali marble forms the central pane, the grains imagine a beautiful landscape scene. The black lacquer on the main frame and cinnabar is used on the perimeter panels.

Double frame with carved panel design may have been popular during the middle to late Ming Dynasty. The format and decoration style show it may be from the south-east of Shanxi province. This piece has similar features to a cinnabar frame and green stone table screen within this collection.

261

046

紫红漆卍字纹框镶绿石座屏

明晚期

高 56 厘米　宽 55.2 厘米　厚 30.5 厘米

中型座屏，薄擦紫漆，纹饰处髹红漆。

屏心绿石，横幅，正背各自成画。正面为绿石常见的奇诡山水状，颇为雄壮，令人遥想起壶口瀑布，中有山崖，其上磨制成八个躬身人形，迤逦而行，若行祷祝之事。背面下方有山峦兀立，余处云气弥漫，山色空蒙。

复框式，横竖枨皆起锐利的剑脊棱，因岁月侵蚀，脊处露白如筋。绦环板皆作长海棠式开光，内透雕卍字纹，并点缀菱花形花朵。其花有一、二、三、四、六朵之别，或单排，或双排，卍字纹局部亦随之变化。

抱鼓墩，下有壸门式亮脚，两足上方雕为卷叶状。对石榴头式站牙，亦随形雕为卷叶状。披水牙板，沿边有红漆卷叶纹，中间为团花，卷叶螺旋相抵如风轮。

此屏薄擦紫漆，民间俗称"羊肝漆"，是北方家具中常见的修饰手法。其造型为典型明中晚期样式，唯与书中同类相比，略为纤薄，或制作稍晚。相似的漆饰见有万历、崇祯款家具，信离此不远。

263

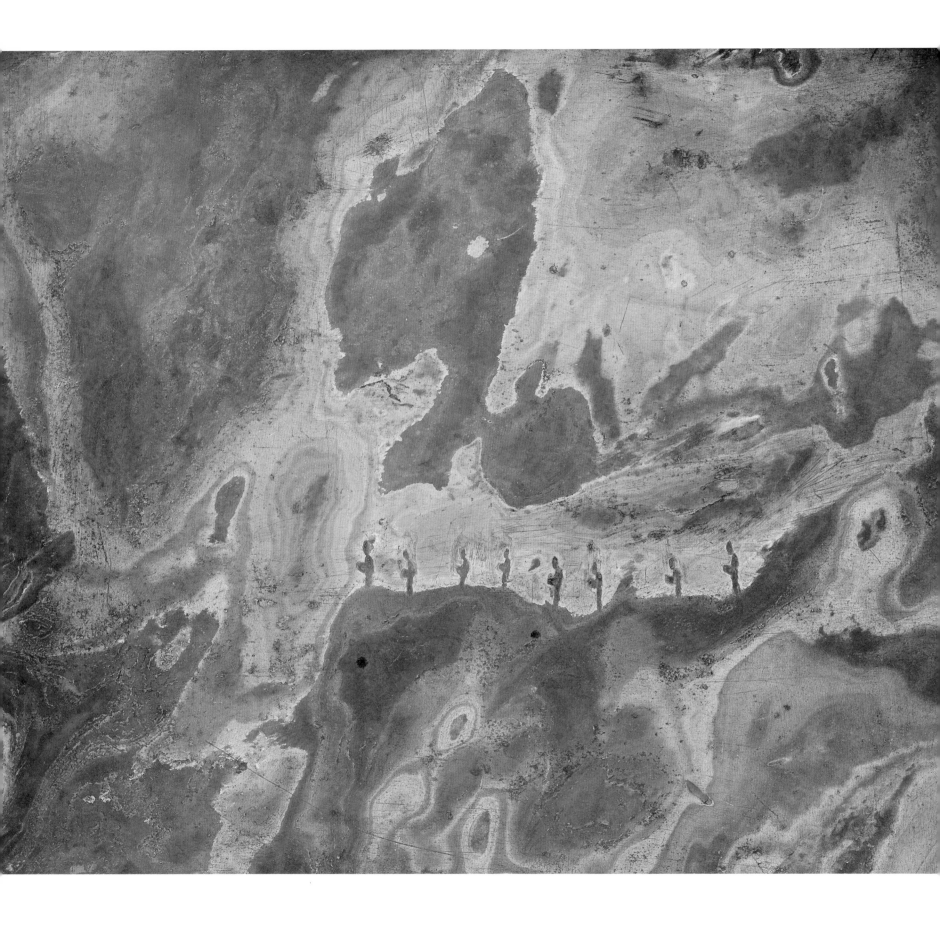

Purple lacquer framed green stone screen

Late Ming Dynasty
H 56CM W 55.2CM D 30.5CM

Medium-sized table screen with thinly rubbed purple lacquer on frame, and cinnabar lacquer on the decoration pattern. Double-frame style, both horizontal and vertical have sharp sword edge. Due to years of wear, the ridges have discoloured appearing light in colour.

This screen is wiped with purple lacquer, commonly known as *yang gan qi*, which is a common decorating method used in northern Chinese furniture. Its shape is typical of the middle or late Ming Dynasty. Similar lacquer decorating techniques can also be seen in Wanli and Chongzhen furniture.

The sash panels are decorated with open carved good fortune symbol and flower patterns. There are one, two, three, four, or six flowers, in either a single or double row.

There is a drum-shaped footing on top of which is carved a leaf shape. The pomegranate head-style standing teeth are carved in the shape of rolled leaves with red lacquer leaf curls along the edge, and a group of flowers in the middle, the spiral leaf curls offset each other like a wind wheel.

Green stone is the heart of the screen and has a special grain resembling a figurine, the rear resembling a landscape.

047

黄花梨花卉纹框镶大理石座屏

清早期
高 95.5 厘米　宽 70 厘米　厚 47.5 厘米

　　此屏高近1米，属于案头陈设座屏中最大的一类。

　　竖屏式，屏心大理石淡青灰色地子，黑褐色伴有黄、白色纹路，波折而多有晕散。双面可观，正面山势雄伟，丛山数重，若夏日草木葱茏；背面山势险峻而居于一角，上方天空若有云气，纹路多白色，若冬日苍山落雪。

　　复框式屏框，内外框沿边起委角线，上、左、右三面一排镶嵌三块绦环板，上开鱼门洞，沿边起阳线装饰。下方三块绦环板上海棠形开光内透雕花卉，双面工，两侧为对称的牡丹纹，中间为梅花纹，皆点缀湖石，布局自然，雕工老辣，镂雕、浮雕、阴刻兼施。披水牙板，壶门式，中间垂牡丹花纹，两头勾卷云纹，为卷叶纹衍成，并伴有努出的草芽，沿边起圆阳线，曲线流畅。壶瓶式站牙，透雕缠枝莲纹，雕工与绦环板的稍异，较为规整，物象落落大方。屏墩抱鼓式，鼓面雕葵花纹，清代《圆明园内工则例》称之为"蕖花瓣"，其形作花瓣两层，各层瓣瓣相叠，形成水涡般动势。鼓下出头雕作卷云纹，如建筑斗拱之麻叶头。

　　屏座下附有"亚"字形地栿，可拆卸，王世襄先生旧藏黄花梨镶石座屏亦见有这种地栿，其他座屏上较少见到。

　　此屏出自北京，早年经北京瀚海拍卖售出，也许出自北京官宦人家。其制式规整，法度严谨，如同缩小的照壁，与常见的文人案头陈设不同，当陈设大案之上，是厅堂陈设用具。大理石纹路正反为夏、冬二景，可因季节不同变化而调换。

HUANGHUALI FRAMED DALI MARBLE TABLE SCREEN

EARLY QING DYNASTY
H 95.5CM W 70CM D 47.5CM

This table screen is almost one meter tall, being a very large size for this genre.

The creamy-white marble with brown and dark grey streaks form a natural design of layered mountainous boulders, mounted in a *huanghuali* frame. The grains and colour of the *dali* marble may appear differently with changes of season.

This piece was sold by Beijing Hanhai Auction Co., Ltd. around 2000. The large size and specific format are different from table screens normally seen. Considering this and the quality, techniques, material and large size, this piece may have originally belonged to a high rank aristocratic family in Beijing.

The size and weight of the marble calls for strong framing which is provided by large double beam surround with openwork to lighten the weight and enhance the beauty of an otherwise heavy look, supported on a removable H-form trestle base.

繁华竞逐

Times of Chasing Prosperity

048
铁梨木
镶白石彩绘人物故事图插屏
清中晚期
高 45 厘米　宽 38 厘米　厚 13.3 厘米

白石屏心，彩绘人物故事装饰，内容为昭君出塞，烟熏状皮壳，甚觉古旧。图绘一着汉装牵马侍从半蹲于地，左腿半屈，供昭君踏蹬上马。昭君身姿窈窕，面带凄容，头戴貂尾，插双雉尾，怀抱琵琶，抬腿踩在侍从腿面。后有侍女，手持彩旗。一旁为络腮胡状番将，面露喜容，亦戴貂尾，插雉尾，手持折扇，回首目视昭君，身躯半躬，态度甚恭。侍从作番人打扮，左手牵马，右手高举，趾高气扬。屋内三个仕女探首窥看，院内两个小童贴耳私语。昭君出塞是汉时故事，以其为主题的画作，宋时开始流行，各代不衰，亦是民间喜闻乐见的装饰题材，版画、年画等工艺美术中时有所见。明清时，人物形象已与历史画不同，多作戏曲人物装扮，动作丰富，表情生动，富有装饰性。屏背为墨笔行书唐王建《十五夜望月》："中庭地白树栖鸦，冷露无声湿桂花。今夜月明人尽望，不知秋思在谁家。"

铁梨木框、座，边框平整，内侧起线装饰。屏下设绦环板，上有笔管式开光，亦沿边起线。刀牙式披水牙板，壶瓶式站牙，长方屏墩，下挖亮脚。

白石彩绘的座屏，多见于山西地区，尤以黄花梨小灯屏最为多见，多绘人物故事、山水图案。与此比例造型相近的座屏也时有所见，以铁梨木制者为多。从人物绘制风格看，属清中晚期制品，简单而趋于僵化的造型，也显示为这一时期。

Tieli framed *Baishi* table screen

Middle to Late Qing Dynasty
H 45cm   W 38cm   D 13.3cm

Table screen featuring a legendary story painted on white marble. This type of table screen is often seen in the Shanxi area, particular with *Huanghuali* framing. This particular table screen is most likely from the Shanxi area. The story is of '*Wang Zhaojun Leaving the Pass Behind*' originating from the Han Dynasty becoming a popular subject in different arts forms since the Song Dynasty. During the Ming and Qing dynasties the depiction of figures invoking this drama became popular. Based on drawing techniques, the colour of the paints and materials used on this piece may have been produce during the middle to late Qing Dynasty.

明崇祯
《绿窗女史》"雁帛"
心远堂藏版

Chongzhen Period, Ming Dynasty
The Story for Ladies
Collection of Xinyuan Tang

277

049

黄花梨
镶石八仙庆寿图插屏

清早中期
高 48.5 厘米　宽 45.5 厘米　厚 20 厘米

　　插屏屏心为一种赭色石质，间有红褐色，质地较软，有行家将之归为"滑石"，尚待研究。其上高浮雕图案装饰，正面为八仙庆寿图，群山耸立，海涛翻涌，红日层云，古松掩映殿宇，松鹤出没，一架虹桥横跨，其上八仙各持法器而行，半空中有仙女捧寿桃驾云而来，其前有二仙一立一坐，作迎接状。背面为山水图案。图案造型稚拙，雕琢爽利干脆，层次丰富，富装饰性。

　　屏框、座黄花梨，皮壳干洁。绦环板上开两个鱼门洞式开光。下设壸门式牙板。立柱平素，变体螭龙式站牙，有角，身躯内卷，花叶形尾抵在立柱上。屏墩较简，转折处做成连弧形造型，下有小足撇出。

　　庆寿主题是座屏中较常见的一类，或为寿诞时馈赠之物，日常摆设厅堂、卧室，以求吉祥寓意。

279

HUANGHUALI AND STONE TABLE SCREEN

EARLY TO MIDDLE QING DYNASTY
H 48.5CM   W 45.5CM   D 20CM

Red and brownish stone with light texture vividly portraying a multi-layered carved scenes on both obverse and reverse sides of the screen. Displayed are highly skilful carving techniques on a clean *huanghuali* frame and stand. Two open carvings are on the panel below the screen with violin shaped spandrels on the sides seated on the base stand. The relief caving depicts the Eight Immortals. The immortality and longevity were subjects commonly used on table screens. May have been used as gifts for auspicious purposes.

050

黑红漆框
镶绿石鸟兽纹座屏

明晚期

高 49.5 厘米　宽 46.5 厘米　厚 28 厘米

　　屏心绿石甚老旧，有石皮。制者利用石皮磨制成物象，山崖上下各立一兽，似猫又似虎，尾巴高举，神态活泼，空中有三鸟盘旋，似在反映鸟兽戏耍的场景。物象仅具轮廓，周匝磨低，类竹刻之留青效果。

　　框、座为软木制，擦黑漆，开光擦红漆。屏墩甚为精彩：屏的竖框至屏墩处变为倒生的草芽，伸出两茎卷草纹，交叉后向两侧袅绕蔓延，以浮雕兼镂雕手法塑造成波折舒卷的草叶，如绸带飘舞，繁而不乱，极有生命力，并盘托两球，形成屏墩。球若日之初升，卷草若云霞相拱托。屏下牙板亦是边缘雕作自屏墩蜿蜒而出的卷草叶纹，花叶延伸为阳线，至中间变为交叉的卷叶，形成壸门式。

　　屏框样式为典型山西风格，石片图案拙朴天真，与江南细腻文雅风格迥异。与此例造型、纹饰相近的座屏，黄花梨制者亦有所见。

　　将石板磨制出鸟兽纹，尤其是猫虎纹，是绿石屏中常见的装饰手法，较经典的此类屏还可举观复博物馆所藏黄花梨螭龙纹框绿石兽纹插屏，屏心石板磨制为一兽，类虎又类猫，惟妙惟肖。

明
赵汝殷
《风林群虎图》卷局部
台北故宫博物院藏

Ming Dynasty
Zhao Ruyin
Part of Handscroll of *Tigers in the Forest*
The Palace Museum, Taipei

BLACK AND CINNABAR LACQUER FRAMED GREEN STONE TABLE SCREEN

LATE MING DYNASTY
H 49.5CM W 46.5CM D 28CM

The ink grey, brown colour stone has similar texture as green stone. Mysterious animal creatures occupy the main scene.

The frame and base made from softwood with black and cinnabar lacquer, decorated with scrolling flower pattern. The decoration pattern, lacquer techniques and format of this piece indicated it may be from Shanxi province. The base is decorated with remarkably ornate spandrels featuring open carved scrolling design on top of intensely detailed scrolled base. The central feature of the base is a pair of spheres appearing to be enmeshed in a complex weave of scrolling grass.

051

紫漆花卉纹框绿石人物故事图座屏

明中晚期

高 41.2 厘米　宽 51.2 厘米　厚 21 厘米

绿石屏心，较扁长，如展开的画卷，质地细腻，以雕版般的手法浅浮雕人物故事图，形象古拙生动。内容为唐代历史演义故事"三跳涧"（亦称三跳虹霓涧），叙秦王李世民与刘武周对战美良川（今山西闻喜县南），被刘的部将尉迟敬德追击至虹霓涧，大将秦叔宝赶来救驾，三者相继跃虹霓涧而过，故名"三跳"。画面中座下马即将就岸，回首观看者为秦王李世民，他双手紧握缰绳，紧张的心情还未舒缓，头上翻滚云雾中金龙乍现，真龙天子自有神灵护佑。涧旁持鞭勒马将跃的虬髯将军为尉迟敬德，山后转出持锏追赶者为秦叔宝。屏心定格于最紧张的跃涧瞬间，涧水声、马蹄声甚至背后战场厮杀声如在耳边，世民之机智，敬德之勇猛，叔宝之赤忠，跃然画面。

三跳涧见于《唐书志传通俗演义》《隋唐志传通俗演义》《说唐传演义全传》《大唐秦王词话》等传奇小说、词话，内容大同小异。创作时应该参照了刘备的卢马跃檀溪的掌故，但内容更加丰富，作为仁君贤臣故事，传播甚广。屏心稿本当出自绣像版画，诸如明万历初版《唐书志传题评》中即有类似插图。唯需指出的是：屏风上的人马形象与云纹、龙纹等尚有元代特征，似乎所本更早。如元刊本《新刊全相秦并六国平话》版画中的人马形象，就与之相近。而屏心之构图，就明显与元刊本《新刊全相平话三国志》中刘备跃檀溪版画有继承关系，可见它们趋近的风格。就此来看，屏心所本稿本也许在明中期或以前。

屏框软木胎，擦羊肝紫色薄漆，复框式，框看面起剑脊棱。复框间设绦环板，剑环式开光内透雕卷草纹、缠枝花卉纹。屏下设窄秀牙板，牙头处为相抵的卷草叶。站牙较简，亦雕饰卷草纹。抱鼓式屏墩，下挖小壶门式亮脚。

以屏的花卉纹样式和漆质来看，应是明晚期制品，当在明万历左右，不排除屏心石板雕刻更早。屏心绿石石质比常见的绿石更细腻，尚需进一步确认。明末清初时刀马人物图案盛行，但座屏屏心上的人物题材以神佛、文士、仕女等为多，刀马人物者少见。三跳涧故事传播虽广，应用于雕刻装饰者较少。

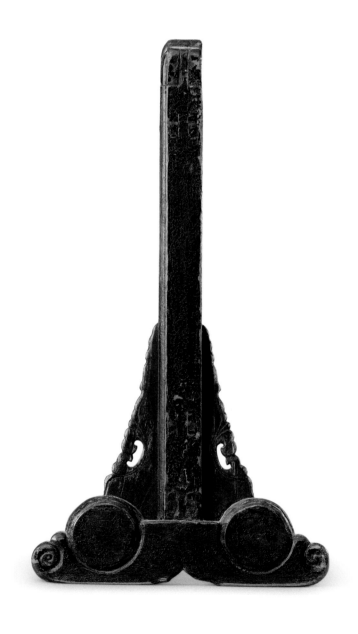

Purple Lacquer Framed Green Stone Table Screen

Middle to Late Ming Dynasty
H 41.2CM W 51.2CM D 21CM

Relief carving technique employed on the marble screen depicts a war story of the emperor Taizong of Tang (598-649 CE). The techniques used on human figures, clouds and dragons share similar characteristics from Yuan Dynasty woodblock prints. In general, the figures carved would be of selected religious deities, scholars or court ladies, warriors are not often portrayed as subjects from this story making this screen somewhat unusual.

The body of the frame and base are made from soft wood coated with purple lacquer. The flower pattern and lacquer indicated this piece may be from the Wanli period of the Ming Dynasty.

052
绿石
人物故事图插屏

清咸丰
高25厘米　宽32厘米　厚8厘米

插屏通体以绿石制成，质地近端石，由屏座、绦环板、屏板三部分四个构件组成。

正面屏心落膛，并雕刻回纹为子框。屏心以线刻饰人物故事图，如同版画，敞厅上夫妻对坐，男主人手捻胡须，笑逐颜开；女主人高髻，一手持花状物，回顾男主人。庭院中，两个仕女走过，前者持灯，显示这是夜晚场景。院西为园林，隔墙可见攒尖亭子，花叶扶疏，墙上长八方形洞门内走来两人，前者面相滑稽，似装扮好的丑角。院东一棵巨松挺立，将画面分割开来，东侧为二层阁楼，十数个巨瓮以一字长蛇式自楼下排至庭院，其上一人身着劲装，两手伸出维持平衡，踏瓮口疾驰而来，乃是表演杂技者。此处所表现的是主人夫妇家中夜观戏曲、杂技的场景。苏轼诗有"他年汝曹筼满床，中夜起舞踏破瓮"句，既发人生之慨，又寄情于后人，与此图内容颇契。整个画面图案较琐碎，下刀稍潦草，为清晚期意趣。画面一角刻印"澹人"。清初高士奇字澹人，但与此屏时代不符。

屏后刻工整的楷书四行："咸丰九年岁在己未孟陬月寿阳祁寯藻镌刻并书。"刻印文"寯藻""实父"。刻制甚精，点画用心，将祁寯藻的清代馆阁体书法特征充分显现。

立柱、站牙、屏墩以一块料雕刻而成，站牙雕为拐子纹。立柱内侧开槽，槽下段为内宽外窄的燕尾式槽，绦环板相应地出燕尾榫，沿槽口自上而下装入，组为一体。绦环板上有落膛笔管式开光装饰。

祁寯藻（1793~1866年）为清晚期名臣，三代帝师，咸丰九年（1859年）时，年过花甲的祁寯藻已致仕六年。正月（即孟陬月），新年伊始，祁翁制此屏，虽然画面表现的是家庭和睦、幸福欢乐的场景，右上角的"澹人"也许是他自况的一方闲印，然从背面工整端庄的款刻就可看出，这样一位忠清亮直、爱国忧民的士大夫，面对内忧外患的局面，不可能淡泊宁静、置身事外。刻制此屏的三年后，同治元年（1862年），年迈的祁寯藻复出，教授年幼的同治皇帝读书，竭诚进讲。同治四年（1865年）因病再次致仕，翌年逝世。

Green stone table screen

Xianfeng Period, Qing Dynasty
H 25cm W 32cm D 8cm

This table screen, panel, base and foot are all made from green stone. Notable is the footing post and spandrel are all hewn from a single piece of green stone. The texture of the stone is close to Duan stone in appearance and quality. The subject of the screen is a densely populated scene of the master being entertained by actors and acrobats while being served by domestic staff.

On reverse side the regular scripts indicated in the colophon written by a famous late Qing officer, Qi Junzao, who was tutor to three emperors. The colophon indicated the date of this piece was made at the beginning of a new year and ordered by Qi Junzao himself and therefore securely dates the piece as well as its original owner.

053

紫檀嵌银丝框
镶青玉云蝠纹座屏

清中期
高 19.5 厘米　宽 21 厘米　厚 8 厘米

屏心玉板，青玉质，雕云蝠捧寿纹，浮雕、阴刻、线刻并施，布局随意，呈现云气弥漫的效果。蝙蝠刻画较为简练，尖嘴，圆眼，桃形耳，头与身躯连为一体，呈枣核形，卷珠状尾，翅如鱼鳍，憨态可爱。背面镶紫檀板，托住玉屏心。

紫檀框、座，满嵌银丝为勾云纹装饰，图案左右对称，中间拼合为兽面纹。绦环板落膛起鼓，起鼓处嵌银丝。披水牙板，轮廓随勾云纹锼挖。屏墩下端内卷如足，侧面两端外撇呈八字形。

屏墩上各边均有规律的后补榫眼，屏框下部空出部分不嵌银丝，显示原应有站牙，但查验实物，几不辨安装站牙的旧痕，概丢失甚早，仅填补榫眼。此屏显然属宫廷家具风格，这种丢失构件后简单修补的原因众多，或是起初即因皇帝不满意站牙，直接摘去，从清宫造办处档案中可以见到很多关于修改家具的记载，也有因乾隆以后国力渐衰，造办处制作水平日下，损坏后难以补配到位，一去了事。

清代宫廷屏具的制作，造型丰富，不再拘泥于常见的传统样式，如此件侧面外撇的屏墩做法，都是当时常见的新变化。此屏可充灯屏、砚屏之用，陈设案头，灯光下银丝闪闪发亮，屏心蝠纹或隐或现，周匝云气变化万千。

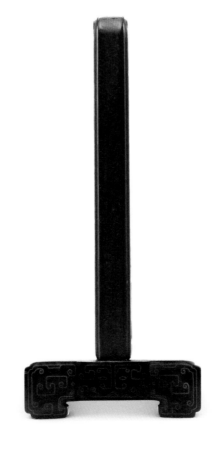

清
徐枚
《松鹤图》轴局部
私人藏

Qing Dynasty
Xu Mei
Part of Hangingscroll of *Crane and Pine Tree*
The Private Collection

Red sandalwood and silver thread framed jade table screen

Middle Qing Dynasty
H 19.5CM W 21CM D 8CM

The central focus of the jade panel is a carved longevity character recessed, surround by clouds and bats by using relief carved techniques. The reverse side of jade panel covered by a red sandalwood panel to protect the precious delicate jade.

The frame and stand fully covered by silver thread with scrolling cloud pattern on the side and mysterious animal creatures in the middle. The materials, decorative patterns and techniques indicate this piece is court furniture. This piece may have become availability as the late Qing court, from time to time, experienced financial problems and had been known to sell off precious items to raise funds.

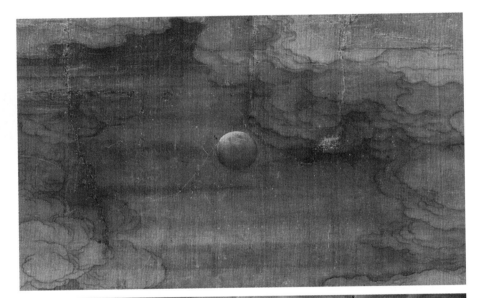

054

花梨木
镶染牙婴戏图小插屏

清晚期
高 18.5 厘米　宽 13 厘米　厚 7 厘米

小型插屏，亦可为灯屏之用。

屏心嵌一块象牙，镂雕球路纹锦地，其上饰婴戏图。图中有红袍童子头戴金冠，右手高持梅枝，左手指日，为喜上眉梢、指日高升、加官晋爵之寓意。旁有童子持挑杆，悬磬，寓意吉庆有余。又有叠罗汉者、奔跑者，共计五个孩童，合五子登科之意。两旁点缀湖石、松树，亦有长寿、喜庆的含义。背面观之，只见物象轮廓，若置灯旁，影影绰绰，别有趣味。

屏框、座花梨木制成，框上以象牙嵌饰一周回纹，绦环板亦为象牙。绦环板、披水牙板、站牙皆雕西番莲纹。

此屏小巧可爱，制作亦精细。这种嵌牙座屏多为清晚期广东制品，行销欧洲，装饰多为喜闻乐见的吉祥图案，如此件袖珍者较为少见。

Rosewood and ivory table screen

Late Qing Dynasty
H 18.5cm W 13cm D 7cm

Small and petit table screen used for protecting a candle flame from the wind. This remarkable screen is carved from one single piece of ivory. The foreground characters and scene is carved in relief whilst the white background is open carving. The foreground is artistically coloured with dyes which have retained their vibrancy over the years.

The frame inlay with ivory with rectangular spiral pattern. The panel below the frame is open carve in lotus scroll pattern.

Carved with a lively design of five boys playfully fighting for a helmet, the scene represents the wish for sons to attain 'first place' in the civil service examinations. The rocks, pine tree on the background carry auspicious meaning with good wishes of the main theme.

This type of table screen made from Guangdong province during the late Qing Dynasty for export European markets.

055

龙泉窑青釉
鱼跃图诗文砚屏

明中晚期
高 16 厘米　宽 16.7 厘米　厚 5 厘米

　　龙泉窑青釉小屏，仅半尺许，为案头所用砚屏，除足底露胎外，通体施肥厚的梅子青釉，釉色青翠欲滴，布满开片。

　　屏心一面饰鱼跃图，刻绘甚浅，开片下隐约可见，波涛中鲤鱼腾跃，空中祥云托日，为常见的鱼跃龙门、指日高升寓意。另一面凸刻楷书诗文，为"万物静观皆自得，四时佳兴与人同"，施釉后形成白色出筋效果，落款"来清斋"。诗文出自宋大儒程颢《秋日偶成》，来清斋不知其人。书法字形饱满，笔画圆润外拓，典型颜体一路。

　　复框式，白色出筋凸起如同木构的横竖枨，上方并列三个开光，中间为如意头形，两边为剑环式，两竖侧为剑环式开光，亦皆出筋，模拟木构的皮条线。诗文屏心一侧下方设两个小管，为插笔之用。屏座呈梯形，中段挖空，余两头为足。站牙略呈石榴头状。

　　龙泉窑为发轫于晚唐五代，造极于宋元明时的南方大窑口，品类极为丰富，行销天下，是极为重要的青瓷产地。砚屏是其中较为特殊的品种，样式大同小异，纹饰以魁星点斗、鱼跃龙门、麒麟等多见，皆是与古代科举考试相关的吉祥寓意。所见此类砚屏以明中晚期制者为多，有外销日本，近年时见回流者。

　　小尺寸的屏多被笼统地称为砚屏，此例有插笔的功能，故又可称为"笔屏"。

Longquan celadon ware table screen

Middle to Late Ming Dynasty
h 16cm w 16.7cm d 5cm

This longquan celadon table screen is covered with a rich green glaze, and ice cracking effect throughout. One side, is a depiction of fish jumping out of water, which is a common symbol of success. The other side is a poem in rectangular script, the poem was written by Cheng Hao (1032-1085ce). Double frame structure, light coloured lines give the impression of two vertical pillars supporting the top beams. The vertical pillars resembling the hinges of a door. The top panel has three carved openings. The side panels with one opening each similar in shape to the top pillar's openings. The surroundings of the openings and the calligraphy are lighter in colour similar to the pillars accentuating and highlighting these embellishments. At the base of the vertical, narrow, pillars are brush receptacles.

Depending on different functional requirements, celadon ware table screens have a rich verity of types. 'Yan ping' is one type within these groups, the piece discusses here was probably made in middle to late Ming Dynasty, some of which were export to Japan during the period.

056

龙泉窑青釉
荷花双鸟图砚屏

明中晚期

高 13.5 厘米　宽 12 厘米　厚 3.5 厘米

　　龙泉窑青釉小屏除足底露胎外，通体施肥厚的淡青釉，图案峰棱处出筋，呈月白色，满布开片。

　　屏心为荷花双鸟图，模压成型，图案简略，下方略呈水波纹，有荷花自右下方向左上方袭绕，花头丰硕，并将画面分割为两个部分，两鸟一上一下呼应，上方还有弯月如钩。

　　复框式，上方三格，两旁为卍字纹，中间如意形开光。竖向两侧亦为如意形开光。底下通长一格，光素无纹。

　　站牙小巧，与底座连为一体，底座出小撇足，较为含混，但仍可辨为如意式足。

　　屏背光素，底座上设两小管，为插笔之用。

LONGQUAN CELADON WARE TABLE SCREEN

MIDDLE TO LATE MING DYNASTY

H 13.2CM　W 12CM　D 3.5CM

This celadon ware table screen is fully covered with a greenish glaze except for the margins on the base footing. The glaze is enhanced with the ice cracks effect throughout. Double frame configuration, three panels on the top section, the middle panel depicting auspicious 'ruyi' (as you wish) sign. The sides panels depict auspicious symbols for good fortune with oblong shape open work on the bottom panel without enhancements. The main central panel has lotus flower, birds and crescent moon embellishments. Petite footings are completed as visually symmetric by a connecting apron. The reverse side is plain without decoration except for two ink brush receptacles.

057

龙泉窑青釉
麒麟纹砚屏

明中晚期
高 16 厘米　宽 16.7 厘米　厚 5 厘米

　　龙泉窑青釉小屏，仅半尺许，为案头所用砚屏，除足底露胎外，通体施肥厚的粉青釉，图案峰棱处出筋，呈月白色，局部有大开片。

　　屏心正面为麒麟望日图案，模压成型，因釉层颇厚，形态浑朴大气，别有一番风味。麒麟巨首，口微张，回首望日，身作蹲踞状，前腿一腿伫立，一腿提起，脚旁有草叶点缀，尾翼翘起与腿足呼应，颇为活泼。空中祥云托日。屏背以寥寥几笔篦划山脉图案。

　　边框较宽，中间打洼，内外缘凸起皮条线，上转角处做成委角。屏墩做成共身双狮纹，喜气洋洋，趴附于屏下，这种狮形坐墩的形象，可以上溯到五代时期，如《重屏会棋图》屏中之山字屏风，即为狮形坐墩。站牙为简化的祥云托日形，披水牙板作壸门状。

　　屏下有与之一体稍成的四面平式小几，尚为宋元形制，带委角，缩面起平线，肩部浑圆，牙板与腿足做成大弧度，沿边有阴线，足端收细落地。屏背设两小管于几座上，为插笔之用。

　　砚屏之制可追溯至北宋时期，此屏复合屏与几案造型，延续宋元制式。

Longquan celadon ware table screen

Middle to Late Ming Dynasty
H 16cm　W 16.7cm　D 5cm

This longquan celadon table screen is covered in a rich green glaze, with ice cracking effect throughout. Wider frame with chamfered shoulders on the two top corners encases a landscape scene. The front depicts an auspicious animal Qilin which is turning its head back look towards the sun which is situated in the top right-hand corner of the screen. The Qilin is seated on the lower margin of the screen which is finished with a water-wave effect.

This screen, appearing heavy in style, is supported by two lions, as footings, the lions are standing on a base pedestal. The pedestal is surrounded by a trench effect before the apron cascading into an arched finish. The whole screen is supported on four substantial unglazed feet. This type of design had its origins in the Song and Yuan dynasties.

058

石立峰山子式屏

明或清

高 15 厘米 宽 11.3 厘米

石青黑色，为竖屏式，若巨嶂耸立。

其形长方，浑然一体，不见斧斤，表面自然起伏状，略成沟壑，有核桃皮式肌理，石质特征不明显，略近英石。中涵一孔，似月似天。

赏石文化滥觞于战汉时期的仙山崇拜，至唐时已经明确成为重要的审美客体，文人笔下已经有以赏石为主题者，有白居易《太湖石记》、吴融《太湖石歌》等脍炙人口的名篇。宋时在米芾等文人推动下，赏石、爱石蔚然成风，尤其是案头陈设赏石盛于一时，有了《素园石谱》等专著。至明清时，则成为文人书斋常有的陈设。

赏石是最为独特的审美客体，源自天然，又别于天然，取其形，又味其形体之外，加以文化的、历史的、艺术的甚至个体的因素，人人不同，时时不同。赏石质地、大小、形体各异，其中独有一类，其形扁薄挺立，可充石屏之用。紫禁城御花园钦安殿东侧小门所对的灵璧石山子式屏，作片状兀立，正面波涛拥簇，底承汉白玉蟠龙纹座，与此例一大一小，可对比品鉴。《天水冰山录》记严嵩有灵璧石屏八座，应也是此类。然立峰山子式屏存世者极少，数年来只见寥寥几例。

MOUNTAIN-FORM'S SCHOLAR'S ROCK

MING OR QING DYNASTY
H 15CM W 11.3CM

Rock of undetermined natural material, rectangular in shape resting on its shortest side in a vertical position. The stone has a rough, uneven surface possibly formed by volcanic action, the appearance of which may be described as brutal and fierce given the surface texture. The focal point is an opening in the middle of the object vaguely resembling a contorted mouth.

Scholar's rocks or 'spirit stone'– ornate natural rock formations that have been prized for their aesthetic properties by Chinese connoisseurs for over a thousand years. Appreciated for form, colour and texture, unusual rocks represented a microcosm of the universe that Chinese scholars could meditate upon in their own studios and gardens.

059

陶座屏模型

宋至元
高 14.6 厘米　宽 8.2 厘米　厚 3 厘米

陶制座屏模型，竖屏式，深色阔边，上方有委角，下方设一体的梯形屏墩。造型虽简，却为研究早期屏风造型的重要资料。

屏风之制甚古，自诞生至今数千年，虽有各种发展衍变，但基本形式还是一直保留下来，如此例造型的座屏，在西汉墓葬所见屏风模型已是如此。此例从造型和比例看，应是宋、辽、金、元时期落地大屏风的模型。

POTTERY TABLE SCREEN

SONG TO YUAN DYNASTY
H 14.6CM　W 8.2CM　D 3CM

Rectangular shaped table screen made of pottery. A glazed surround forms a heavy frame with chamfered corners on the top border. The frame encapsulates an unglazed section which forms the principal feature of the table screen. The footings are in similar material with the main body giving a consistency in colour and texture. The pottery feature, being unglazed, reveals an interesting pattern of various mineralisation the actions of which may have been enhanced over time considering the antiquity of the piece.

060

黄花梨
灵芝纹屏座残件

明末清初

残高 16.8 厘米　宽 25 厘米　厚 17 厘米

　　屏座取宽厚黄花梨板，一木雕出立柱和其间的花板。立柱上端外撇，这种做法在河北赞皇西高村赵郡李氏家族墓所出的北朝青釉笔架上已有所见，惜柱头处已残，不知造型为何。立柱间花板透雕灵芝纹，构图大气，雕琢古朴生动，饶有明风。

　　上部残失，不知何种结构，从现有花板上可发现原有嵌物的痕迹，为"八"字形，推断或为鼎形"寿"字下端的两笔，"寿"字概以玉或其他珍贵材料制成，形成灵芝拱寿的图案。

　　站牙、披水牙板已失。抱鼓式屏墩，鼓侧浮雕蕖花瓣，雕刻延续古朴风格。现披水牙板的槽口已延至屏墩下端，原屏墩应更高，抱鼓下原应有地栿之类的结构座式造型。

　　此座屏残件保持黄花梨老辣原皮壳，兼以别致的造型、大气的用料、生动的雕刻，虽然残失较多，但风华未失，令人不忍释手，胜出普通座屏甚多。

HUANGHUALI STAND

LATE MING TO EARLY QING DYNASTY
H 16.8CM W 25CM D 17CM

The stand made from thick *huanghuali* timber, with side posts and open carved panels made from a signal piece of this fine wood. The top end of the posts are flared slightly outwards, similar to posts seen on archaeology discoveries from the Li family tomb in Zanhuang Xigao village in Hebei province. Unfortunately, the head of the posts of this piece are lost, and therefore create uncertainty as to the original design of the post.

The open carving features the highly auspicious plant *lingzhi*, a pattern commonly seen on wooden carvings from the late Ming Dynasty.

From the remaining sections of this piece, there are traces of inlays which have been lost. Despite this there are signs that the character missing could be that of *shou* (longevity) which matched with the auspicious *lingzhi* pattern. Lost also are the spandrels and apron, the base footings displays a round drum shaped carving of flower petals on the outer and inner sides of the drum.

This piece retains the colourful character of *huanghuali* and to be crafted out of one piece of wood maintains clear evidence of the skills and artistry of this remarkable feat of craftsmanship.

跋

蕉园主人

人生过半，经历和感悟良多。再回首，所受世间眷顾，远不是一句感恩可以道尽。一路走来，甘苦自知，更多的，是快乐。而快乐的源头，一半来自朋友，另一半也来自朝夕相处的"朋友"，只不过这些"朋友"是无声的，那便是我的收藏。

生长于古都北京，老城掌故、古物风华，耳闻目睹，但那时候，还不能理解它的美妙，只是有一种神秘的力量，一直吸引我、召唤我。我18岁的第一份工作，就进入了琉璃厂"怡仿斋"画廊。"怡仿斋"是李苦禅先生创办，画廊多种经营，有文房四宝、工艺品，当然更多是书画。我分配在书画组，每天挂上搬下，接待的多半是外国人。虽然只有短短一年，但对收藏有了直观的印象，从此埋下未来追求的种子。

伴时代腾飞，经过自己的努力，有了一些积蓄，便更多地涉猎于收藏。最初，也是由熟悉的近现代书画入手。我的第一件藏品，是齐白石画的"葫芦"，老先生笔下的情趣让人生爱，后来一直是我收藏的重点。黄胄先生的才情也令我景仰，每遇精品，欲罢不能，日积月累而成收藏系列。其他如傅抱石、徐悲鸿、张大千、于非闇、李可染、吴冠中、王雪涛等大家，乃至更早的任伯年等，或多或少，都有收获。当年还没有形成今天这般的收藏热潮，挑选精品力作的机会多，本着宁缺毋滥的原则，每一件都来历清晰、传承有序，成为我不舍的陪伴。

当墙上挂上这些悦目的画作，美意至此，生活空间应该也相辅相成，一般"世俗"物件，好像不能满足我的要求，似乎还有更玄妙的东西要来补足。是什么呢？古意。古意不仅能让我们生活在历史中，还使生命有了更久远的寄托。兴之所至，我又以更大的热情，投入到"古物"的搜集中。

尤其是明式家具和文房器物，上通天地精神，下抱文人情怀，将中国传统思想、美学、趣味，推向后世难以企及的高度，把玩再三，心中的暗喜，无以言表。有一段时期，几乎每天都要去吕家营旧物市场、高碑店家具一条街，眼见得这一行聚散如烟云，逐渐尘埃落定，各有其所，而我也在这过程中，锻炼了眼力，提高了眼界，收获了爱物。

最让我感慨，也最让我珍视的，是从收藏的经历中，结交了不少终身好友。从书画到文房，从行家到专家，从拍卖行到博物馆，因为同好，不同职业、不同背景、不同地方的人，时不时相聚、畅叙、雅集，彼此都抱着旧时明月的襟怀。当然也有争长论短，遇好物而互不谦让，这些经历，使我的生活增添了无尽的乐趣。

真正有着相当年龄的古器物，包括文房家具，它的包浆，代表着历史的温度，为我所得，为我所享，除了美，还是缘。每一件长物，各有各的亮点，素而不简，妍而不媚，巧而不繁，游历于权贵、文人与工匠之间，将自然和俗世融为一体，焉有不爱的道理？但万物终有其归类，世上好物无数，也只能万取一收。与近现代书画收藏一样，我的古物收藏也有一些偏好，大概分类，有以下侧重。

明式家具最为我心仪，还在于它可用、可观、可怡情，与生活息息相关。黄花梨材质上等，又有文人参与制作，美学上的精简呼应了现代感，被世界范围所热捧。能收获一二，付出的不仅是财力，而且是精力，当然还有运气。黄花梨家具品类多，经过这些年不辞辛苦的追寻，也基本齐全，精益求精之下，不乏佳作良器。其他材质，如鸂鶒木、铁梨木，甚至榉木，只要审美达到一定的高度，我同样倾心收入，并不输黄花梨的位置。

座屏也属于家具的一部分，但它的实用功能不强，往往为一般藏家所忽略。正因为"无用"，更显出它的独立本色。难怪古代绘画中的高士不离座右，原来就是"位"的定力。确实，在日常空间中，一旦摆放好座屏，召集周边的杂物，立马提气，氛围显得精神十足。历数元明清各种年代制式，大漆螺钿、紫檀花梨、精瓷美玉，不觉各种佳构达百余，是我相对丰富的收藏体系。尤其屏心的自然石纹，有山水奇观的抽象美，又与我喜爱的纸上云烟联系起来，坐定家中，神游天地间。

其他如赏石、匾额，以及造像、文房小件、陶器等，积年累月，略有浮泛涉猎。

或许从我的收藏，能发觉它们共同的特质，本是关乎日常的点缀，生活的锦上添花，我乐于被美好事物所环绕。因此，由内而外，我也是一个造园爱好者，筑亭、理水、置石、移树、弄花、观鱼，乐在其中，闲看四时之景，与室内收藏隔窗对照，再有几多好友对酒当歌，大有"不知今夕是何年"的感慨。"非典"那年，有幸得到齐白石老先生所题"蔗园"

书额,有不断进步节节高的寓意,"蔗园"便取为堂号,算是生活的归宿。

在我看来,收藏不仅与我个人爱好有关,也能潜移默化,向后辈传递惜物之道,影响他们对传统文化的态度,培养他们的审美和格调。前人之藏也无心,而后人无心焉好之,谓种瓜得瓜,种豆得豆,这是我的期望。

不知不觉浅涉收藏已有二十余年。我们不过是物质的暂时拥有者、保管者、享用者,因此有必要总结和汇报阶段性的经眼,借部分藏品的整理、编辑和出版,表达对祖国优秀传统文化和艺术的敬意,同时感谢这么多年来,各位专家学者、同道好友的帮助和支持,期待未来更好地与人相惜,与物相宜。

感谢邹静之大兄撰写的序言,感谢刘丹大兄的座屏题跋,令本书增色甚多。感谢徐累、张金华、李猛、李捷、张志辉、唐卫国、武建云等"座右工作组"人员的辛苦劳动,跨时五年的工作让我们的友情更加深厚。黄玄龙兄、姚櫚兄等对本书提出的诚恳意见,让我获益匪浅。感谢家人的支持。感谢故宫出版社出色的编辑出版工作。还有很多提供帮助和关心本书的朋友,总是令我受宠若惊,恕不一一,铭记于心。

收藏一门,始于情结,欣于所遇,趣于真伪,识于风格,合于审美,妙于会心,久于懂得,终于岁月,贵于传承,乐于分享。

是为记。

Postscript

The Owner of Zhe-garden

As I have passed the halfway point of my life, I have lived in harmony with the world and kept up with the development of the times. Looking back, I am grateful for the favours I have received from the world, which cannot be fully expressed in words. Along the way, I have experienced both joy and sorrow, but the former has outweighed the latter. And the source of my happiness comes partly from my friends, and partly from my collection, made up of the objects that have been with me day and night, though they are silent.

I was born and raised in Beijing, the ancient capital of China. I was exposed to the stories and legends of the old city from a young age and was fascinated by them and the antiquities surrounding me. In my youth however, I could not fully appreciate their beauty. I simply felt the mysterious power, which seemed to be calling me. At the age of 18, I got my first job at the 'Yifangzhai' gallery of Liulichang, founded by Mr. Li Kuchan. The gallery dealt in a variety of items, including the 'four treasures of study' in ancient China and handicrafts, calligraphy and painting. I was assigned to the calligraphy and painting department, where I displayed and demonstrated works of art for mostly foreign customers. Although I only worked there for a short time, it left me with a deep thirst for collecting. It planted the seeds of my future pursuit, and I began to pursue the idea of collecting.

Thanks to the changing times and through my own efforts, I had accumulated some savings and was able to fully devote myself to collecting. I started with familiar modern and contemporary Chinese paintings and calligraphy. My first collection was a painting of a 'gourd' by Qi Baishi. The charm of Mr. Qi's brushwork fascinated me and has always been a key focus of my collection. I also admire the talent of Mr. Huang Zhou, and whenever I encounter a masterpiece, I cannot resist the temptation to acquire it. Over time, my collection has grown to include other famous Chinese artists such as Fu Baoshi, Xu Beihong, Zhang Daqian, Yu Fei'an, Li Keran, Wu Guanzhong, Wang Xuetao, and even earlier artists like Ren Bonian. Fortunately, during that time, there was not yet a craze for collecting as we have today, and there were many opportunities to select high-quality works. Adhering to the principle of preferring to miss out rather than acquire inferior works, every piece in my collection has a clear origin and provenance, becoming my cherished companions.

As these beautiful paintings hang on the walls, they should complement and enhance the living space. Ordinary 'mundane' objects seem to fall short of my requirements, and it seems that there are more mysterious things to fulfil this objective. Such as antiquities. Antiquities not only allow us to live in history but also provide a sense of connection with a more distant past. With great enthusiasm, I have immersed myself in this endeavour.

Especially in the case of Ming-style furniture and Scholar's studio objects which embodies the spirit of connecting heaven and earth and embrace the sentimentality of the literati. They bring Chinese traditional ideas, aesthetics, and interests to unparalleled heights, and the joy in my heart when I handle them is beyond words. For a period of time, I visited the Lüjiaying Antique Market and Gaobeidian Furniture Street frequently. I saw this industry come together and then disperse like smoke, gradually settling, each with its own place, and in the process, I honed my skills and improved my taste, rewarded by the love of these objects.

What impressed me, and what I treasure the most, is my collecting experiences has allowed me to make many lifelong friends. From connoisseurs to experts, from auction houses to museums, people from different professions, backgrounds, and places gather from time to time, converse, and hold elegant gatherings, all bound by their shared interests. Of course, there are also disagreements and debates over the merits of various items, but these experiences have added infinite joy to my life.

For antique objects that have reached a certain level, including study furniture, their 'patina', represents the warmth of history. They are obtained and enjoyed not only for their beauty, but also for their stories. Each object of antiquity has its own highlights, simple yet elegant, charming yet not flamboyant, intricate yet not excessively ornate. They travel among the nobles, literati, and artisans, blending nature and the secular world into one, making it irresistible to love. Everything has its classification, and although there are countless treasures in the world, only one in ten thousand can be a collectable. Similar to collecting modern and contemporary paintings and calligraphy, my antique collection also has some preferences, which can be roughly categorized as follows.

I am particularly fond of Ming furniture because it is visually appealing, practical, and emotionally satisfying, closely related to daily life. Although *Huanghuali* wood is of superior quality and have literati involvement in its production, its aesthetic simplicity echoes modern sensibilities, making it highly sought after worldwide. To obtain a piece, one must not only have the financial means, but also invest time and energy, as well as some good fortune. After years of hard work, my collection of *Huanghuali* furniture is nearly complete, with constant efforts to improve the quality of the collection. As for other materials, such woods as *Xichi*, *Tieli*, and even elm, as long as the aesthetic standards are high, I am equally devoted to and do not consider inferior to *Huanghuali*.

Screens belong to the category of furniture, but their practical function is minimal, and they are often overlooked. Because some consider them as of little practical value, therefore it is no wonder that the high-ranking people in ancient paintings rarely left their screens. In fact, once the screens are placed in a daily space, they immediately blend into the surrounding items and create an uplifting atmosphere. Therefore, I have gathered various styles from the Yuan, Ming, and Qing dynasties, made of lacquer, mother-of-pearl inlay, red sandalwood, rosewood, delicate porcelain, and jade. The natural stone patterns on the 'screen cores' have an abstract beauty of landscapes, which reminds me of the smoke and clouds on my favourite drawings. Sitting in my home, I can wander in the world of mountains and rivers.

Other items such as decorative stones, plaques, statues, stationery, and pottery have also been collected over the years, although to a lesser extent.

Perhaps from my collection, one can discover their common characteristics – they are all related to daily life's enhancements, adding beauty to my surroundings. I am therefore also a gardening enthusiast, building pavilions, arranging water features, placing stones, moving trees, growing flowers, and raising fish. I enjoy observing the scenery of the four seasons leisurely. Contrasting my indoor collections through the windows, and sharing a few drinks with friends, I can't help but feel the passage of time. In the year of the SARS epidemic, I was fortunate enough to receive a calligraphy plaque Zhe Yuan (Zhe-garden) from the late master Qi Baishi, which symbolized my continuous progress and growth. From then on, Zhe Yuan became the name of my hall, and it has become my home in life.

Time flies, and as we exchange with objects, we unconsciously become collectors for over twenty years. We are only temporary possessors, guardians, and enjoyers of material objects. Therefore, it is necessary to summarize and report on the stages of experience, by organizing, editing, and publishing some of the collections, expressing respect for the excellent traditional culture and art of our country. Thanks for the many experts, scholars, and friends who have helped and supported me over these years. I look forward to cherishing people and things even better in the future.

I'm so grateful to Zou Jingzhi for his foreword and Liu Dan for his postscript, both articles add much to the book. Special thanks to the team for *Seat Right*. They are Xu Lei, Zhang Jinhua, Li Meng, Li Jie, Zhang Zhihui, Tang Weiguo, Wu Jianyun, etc. The five years' work has deepened our friendship. I appreciate Huang Xuanlong and Yao Gang for sincere comments on this book. I would like to thank the Forbidden City Publishing House for its excellent editing and publishing work. Finally, I would extend my gratitude to my family and friends for their support and understanding.

Collecting is about emotions, enjoyment, authenticity, style, aesthetics, appreciation, understanding, and time, ultimately leading to inheritance, and the joy of sharing.

图版索引
Index

001
黑漆薄螺钿孔子观欹器图座屏
Mother-of-pearl inlaid black lacquer table screen
—
62

002
黑漆薄螺钿佛像图座屏
Mother-of-pearl inlaid black lacquer table screen
—
68

003
黑漆薄螺钿雅集图座屏
Mother-of-pearl inlaid black lacquer table screen
—
74

004
黑漆薄螺钿人物故事图座屏
Mother-of-pearl inlaid black lacquer table screen
—
80

005
黑漆薄螺钿三仙贺寿图插屏
Mother-of-pearl inlaid black lacquer table screen
—
88

006

黑漆薄螺钿缠枝花纹框镶红漆诗文座屏

BLACK LACQUER FRAMED INLAID WITH CINNABAR LACQUER TABLE SCREEN

92

007

黄花梨螭龙纹框黑漆刻灰水仙图插屏

HUANGHUALI FRAMED KEHUI BLACK LACQUER TABLE SCREEN

96

008

剔红山水人物图座屏

CARVED CINNABAR LACQUER TABLE SCREEN

102

009

黑漆框剔红云龙纹座屏

BLACK LACQUER FRAMED CARVED CINNABAR LACQUER TABLE SCREEN

110

010

紫漆百宝嵌秋郊饮马图座屏

PURPLE LACQUER INLAID PRECIOUS STONE TABLE SCREEN

114

011

紫檀嵌石、牙二甲传胪图插屏

RED SANDALWOOD INLAID IVORY AND STONE TABLE SCREEN

120

012

黑红漆麒麟纹座人物故事图插屏

BLACK AND CINNABAR LACQUER TABLE SCREEN

126

013

红漆撒螺钿框山水图座屏

CINNABAR LACQUER WITH CRUSHED MOTHER-OF-PEARL FRAMED TABLE SCREEN

130

014

红木框黑漆彩绘博古图座屏

HONGMU FRAMED BLACK LACQUER POLYCHROME TABLE SCREEN

134

015

黑红漆山字式可折叠座屏

BLACK LACQUER FOLDING TABLE SCREEN WITH GEOMETRIC LATTICE OPEN WORK

136

016
黄花梨镶大理石座屏
HUANGHUALI AND DALI MARBLE TABLE SCREEN
142

017
乌木镶大理石座屏
WUMU AND DALI MARBLE TABLE SCREEN
148

018
黄花梨镶绿石座屏
HUANGHUALI AND GREEN STONE TABLE SCREEN
152

019
黄花梨镶白石座屏
HUANGHUALI AND BAISHI TABLE SCREEN
156

020
黑漆框镶大理石座屏
BLACK LACQUER FRAMED DALI MARBLE TABLE SCREEN
158

021
紫檀镶大理石座屏
RED SANDALWOOD AND DALI MARBLE TABLE SCREEN
160

022
黄花梨镶大理石插屏
HUANGHUALI AND DALI MARBLE TABLE SCREEN
164

023
黄花梨镶大理石插屏
HUANGHUALI AND DALI MARBLE TABLE SCREEN
166

024
黄花梨镶绿石插屏
HUANGHUALI AND GREEN STONE TABLE SCREEN
168

025
黄花梨镶大理石插屏
HUANGHUALI AND DALI MARBLE TABLE SCREEN
172

026
黄花梨夔纹框镶大理石座屏
HUANGHUALI FRAMED DALI MARBLE TABLE SCREEN
176

027

黄花梨镶绿石座屏

HUANGHUALI AND GREEN STONE TABLE SCREEN

182

028

黄花梨镶大理石座屏

HUANGHUALI AND DALI MARBLE TABLE SCREEN

186

029

紫檀镶大理石插屏

RED SANDALWOOD AND DALI MARBLE TABLE SCREEN

190

030

黄花梨镶大理石插屏

HUANGHUALI AND DALI MARBLE TABLE SCREEN

198

031

黑漆撒螺钿框镶大理石座屏

BLACK LACQUER MOTHER-OF-PEARL FRAMED DALI MARBLE TABLE SCREEN

200

032

楠木镶大理石座屏

NANMU AND DALI MARBLE TABLE SCREEN

204

033

黄花梨双螭纹框镶绿石插屏

HUANGHUALI FRAMED GREEN STONE TABLE SCREEN

208

034

黄花梨螭龙纹框镶绿石山水人物图座屏

HUANGHUALI FRAMED GREEN STONE TABLE SCREEN

210

035

紫檀嵌玉棂格镶大理石座屏

RED SANDALWOOD, JADE AND DALI MARBLE TABLE SCREEN

216

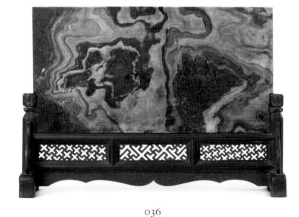

036

红擦漆卍字纹座大理石插屏

CINNABAR LACQUER AND DALI MARBLE TABLE SCREEN

222

037

乌木座大理石插屏

WUMU AND DALI MARBLE TABLE SCREEN

228

038

黄花梨镶大理石插屏

HUANGHUALI AND DALI MARBLE TABLE SCREEN

232

039
紫檀镶大理石插屏
RED SANDALWOOD AND *DALI* MARBLE
TABLE SCREEN
234

040
黄花梨螭龙纹框镶绿石插屏
HUANGHUALI FRAMED GREEN STONE SCREEN
238

041
紫檀镶大理石插屏
RED SANDALWOOD AND *DALI* MARBLE
TABLE SCREEN
240

042
黄花梨镶大理石插屏
HUANGHUALI AND *DALI* MARBLE TABLE SCREEN
244

043
黑红漆镶大理石座屏
BLACK AND CINNABAR LACQUER *DALI* MARBLE
TABLE SCREEN
250

044
红漆框镶绿石座屏
CINNABAR LACQUER FRAMED GREEN STONE
TABLE SCREEN
254

045
黑红漆框镶大理石座屏
BLACK AND CINNABAR LACQUER *DALI* MARBLE
TABLE SCREEN
258

046
紫红漆卍字纹框镶绿石座屏
PURPLE LACQUER FRAMED GREEN STONE SCREEN
262

047
黄花梨花卉纹框镶大理石座屏
HUANGHUALI FRAMED *DALI* MARBLE TABLE SCREEN
266

048

铁梨木镶白石彩绘人物故事图插屏

TIELI FRAMED BAISHI TABLE SCREEN

274

049

黄花梨镶石八仙庆寿图插屏

HUANGHUALI AND STONE TABLE SCREEN

278

050

黑红漆框镶绿石鸟兽纹座屏

BLACK AND CINNABAR LACQUER FRAMED GREEN STONE TABLE SCREEN

282

051

紫漆花卉纹框绿石人物故事图座屏

PURPLE LACQUER FRAMED GREEN STONE TABLE SCREEN

286

052

绿石人物故事图插屏

GREEN STONE TABLE SCREEN

292

053

紫檀嵌银丝框镶青玉云蝠纹座屏

RED SANDALWOOD AND SILVER THREAD FRAMED JADE TABLE SCREEN

296

054

花梨木镶染牙婴戏图小插屏

ROSEWOOD AND IVORY TABLE SCREEN

300

055

龙泉窑青釉鱼跃图诗文砚屏

LONGQUAN CELADON WARE TABLE SCREEN

302

056

龙泉窑青釉荷花双鸟图砚屏

LONGQUAN CELADON WARE TABLE SCREEN

304

057

龙泉窑青釉麒麟纹砚屏

LONGQUAN CELADON WARE TABLE SCREEN

306

058

石立峰山子式屏

MOUNTAIN-FORM'S SCHOLAR'S ROCK

309

059

陶座屏模型

POTTERY TABLE SCREEN

310

060

黄花梨灵芝纹屏座残件

HUANGHUALI STAND

312

附录 | 检测报告

APPENDIX: TEST REPORT

加速器质谱分析结果
Accelerator Mass Spectrometry Result

校准报告
Calibration Report

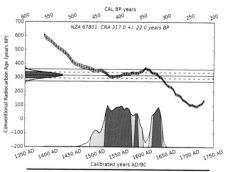

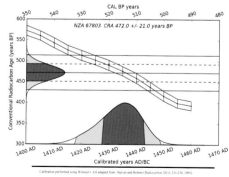